Good Laboratory Practice Regulations

DRUGS AND THE PHARMACEUTICAL SCIENCES

A Series of Textbooks and Monographs

edited by

James Swarbrick
AAI, Inc.
Wilmington, North Carolina

Modern Pharmaceutics: Third Edition, Revised and Expanded, *edited by Gilbert S. Banker and Christopher T. Rhodes*

Pharmaceutical Powder Compaction Technology, *edited by Christer Nyström and Göran Alderborn*

Good Laboratory Practice Regulations

Second Edition, Revised and Expanded

edited by

Sandy Weinberg

Weinberg, Spelton & Sax, Inc.
Boothwyn, Pennsylvania

MARCEL DEKKER, INC.   NEW YORK · BASEL · HONG KONG

Library of Congress Cataloging-in-Publication Data

Good laboratory practice regulations / edited by Sandy Weinberg. —
    2nd. ed., rev. and expanded.
        p.   cm. — (Drugs and the pharmaceutical sciences ; v. 69)
    Previous ed. edited by Allen F. Hirsch.
    Includes bibliographical references and index.
    ISBN 0-8247-9377-3 (alk. paper)
        1. Medical laboratories—Quality control.   2. Medical
    laboratories—Law and legislation.   3. Biological laboratories-
    -Quality control.   4. Biological laboratories—Law and legislation.
    I. Weinberg, Sandy.   II. Series.
    R860.G66   1995
    574'.072—dc20   95-10741
                                                                              CIP

The publisher offers discounts on this book when ordered in bulk quantities. For more information, write to Special Sales/Professional Marketing at the address below.

This book is printed on acid-free paper.

MARCEL DEKKER, INC.
270 Madison Avenue, New York, New York 10016

Current printing (last digit):
10 9 8 7 6 5 4 3 2

PRINTED IN THE UNITED STATES OF AMERICA

PREFACE

Good Laboratory Practice Regulations (GLPs) are promulgated by the Commissioner of the U.S. Food and Drug Administration (FDA) under general authority granted by the Federal Food, Drug and Cosmetic Act. The GLPs were first published in 1976. Final regulations were published in 1978 with an effective date of June 20, 1979. The regulations have not been static; revisions have been issued in 1980, 1987, twice in 1989 and again on September 13, 1991.

The GLPs prescribe standards for the conduct of studies designed to establish the safety of products regulated by the FDA. Sponsors may conduct the studies in their own laboratories or have them performed by a contract laboratory, a university, or some other type of laboratory. The sponsor submits the study reports to the FDA in food and color additive petitions, investigational new drug applications, new drug applications, new animal drug applications, biological product license applications, and other requests for permission to market a product. If the agency accepts that safety and efficacy are established adequately, marketing of the product is permitted.

The foregoing should not imply that the FDA is the sole regulatory body concerned with good practices in the lab. The U.S. Environmental Protection Agency (EPA) has finalized GLPs for both the Toxic Substances Control Act and the Federal Insecticide, Fungicide, and Rodenticide Act. The language of these regulations has been harmonized to the extent possible with the language of the FDA GLPs. The scope of the EPA's regulations affects all types of health and environmental testing including environmental fate testing, ecological effects, residue chemistry, efficacy testing, field studies, and epidemiology studies.

One should not believe that the United States has "cornered the market." GLPs are part of the international marketplace. The U.S. and other members of the 24-nation Organization for Economic Cooperation and Development (OECD) have been involved in extensive international consultations in efforts to bring industrial and environmental chemical programs into harmony.

Analytical labs, while not specifically addressed in the regulations, may find it advantageous to be GLP compliant in environments where GLPs are not a requirement. A thorough discussion of the GLPs must include how the GLPs are monitored.

This second edition of *Good Laboratory Practice Regulations* addresses the above topics as well as discussing all aspects of FDA's GLP regulations and techniques for implementation. These topics are presented within the contexts of a historical perspective and current laboratory automation. Since computer-aided study design, sample selection, data collection, and analysis and storage of information are part-and-parcel of the contemporary laboratory, separate chapters consider the topics of the automated laboratory and computer validation.

Chapter 1 gives a historical perspective of the FDA which focuses on problems and solutions; proposed regulations from the FDA and the EPA; and GLP revisions. Chapter 2 provides a general discussion of all aspects of the FDA's GLP regulations. Chapters 3 and 4 discuss the EPA, foreign country GLP regulations and the EPA Federal Insecticide, Fungicide, and Rodenticide Act (FIFRA) GLPs. Chapter 5 introduces economic behavior as a measure to evaluate and optimize the use of automation and instrumentation technology in regulated laboratories. Chapter 6 addresses computer system validation and how it establishes the credibility of laboratory data and automated procedures. Chapter 7 presents a regulator's perspective of the laboratory inspection process. Chapter 8 presents summary commentary and ventures a forecast as to the future value and effectiveness of the GLPs as robotic labs become more common. An up-to-date bibliography provides useful references for GLPs.

Regardless of the laboratory environment, or location, its position as practitioner or regulator, or the extent of automation employed, several common themes are detected throughout this volume. These are adherence to and adoption of several key provisions of the GLPs, including:

- The creation of a quality assurance unit (QAU) whose function is to inspect and audit nonclinical laboratory safety studies and the accompanying data.
- The appointment of a study director who has ultimate responsibility for the study.
- The need for written protocols and SOPs.
- The requirement to analyze the test and control article/ carrier mixture.
- The necessity to utilize instruments that are maintained, calibrated, and standardized.

Sandy Weinberg

CONTENTS

Chapter 4

Chapter 5

Chapter 6

Chapter 7

Chapter 8

CONTRIBUTORS

JAMES N. BOWER, Ph.D.   Chief Technical Officer, Automated Compliance Systems, Inc., Bridgewater, New Jersey

JOHN J. FITZGERALD, Ph.D.   President, Automated Compliance Systems, Inc., Bridgewater, New Jersey

GEORGE W. JAMES, Ph.D.   Chief, Nonclinical Laboratory Studies Branch, Office of Compliance, Center for Drug Evaluation and Research, U.S. Food and Drug Administration, Rockville, Maryland

CARL R. MORRIS, Ph.D.   President, International Chemical Consultants U.S.A., Inc., Alexandria, Virginia

WENDELL A. PETERSON, J.D.   Director, Quality Assurance, Parke-Davis Pharmaceutical Research Division of Warner-Lambert Company, Ann Arbor, Michigan

FREDERICK G. SNYDER   Director of Quality Assurance, Astra/Merck Group of Merck & Co., Inc., Wayne, Pennsylvania

GARY C. STEIN, Ph.D.   Director of Research and Regulatory Affairs, Weinberg, Spelton & Sax, Inc., Boothwyn, Pennsylvania

BARBARA N. SUTTER   Analytical Quality Resource Coordinator, Analytical Sciences Laboratory, The Dow Chemical Company, Midland, Michigan

JEAN M. TAYLOR, Ph.D.   Retired Team Leader, Division of Toxicology, Center for Food Safety and Applied Nutrition, U.S. Food and Drug Administration, Washington, D.C.

SANDY WEINBERG, Ph.D. President, Weinberg, Spelton & Sax, Inc., Boothwyn, Pennsylvania

Good Laboratory Practice Regulations

Chapter 1

HISTORICAL PERSPECTIVE

Jean M. Taylor, Ph.D.[1]

U.S. Food and Drug Administration, Washington, D.C.

Gary C. Stein, Ph.D.

Weinberg, Spelton & Sax, Inc., Boothwyn, Pennsylvania

THE PROBLEM IN THE 1970s

FDA's Perspective

The Federal Food, Drug and Cosmetic Act (FFDCA) places the responsibility for establishing the safety and efficacy of human and veterinary drugs and devices and the safety of food and color additives on the sponsor of the regulated product. The Public Health Service Act requires that a sponsor establish the safety and efficacy of biological products. These laws place on the Food and Drug Administration (FDA) the responsibility for reviewing the sponsor's test results and determining whether the results establish safety and efficacy of the product. If the agency accepts that safety and efficacy are established adequately, marketing of the product is permitted.

    The types of scientific tests needed to establish safety are dependent on the nature of the regulated product and its proposed use. A product such as a food or color additive will require tests to elucidate the potential of the product to induce adverse acute, subchronic, and chronic

---

[1]Retired.

effects. The safety tests are generally performed in animals and other biological systems. Both the types of tests and the methodology of particular tests have changed over the years with scientific advances in the field of toxicology.

FDA regulations or guidelines prescribe the type of safety tests for a particular product. Sponsors may conduct the studies in their own laboratories or have them performed by a contract laboratory, a university, or some other type of laboratory. The sponsor submits the study reports to the FDA in food and color additive petitions, investigational new drug applications, new drug applications, new animal drug applications, biological product license applications, and other requests for permission to market a product.

FDA scientists evaluate the safety studies to determine whether the results support a conclusion that the product can be used safely. Until the mid-1970s, the underlying assumption in the agency review was that the reports submitted to the agency accurately described study conduct and precisely reported the study data. A suspicion that this assumption was mistaken was raised in the agency's review of studies submitted by a major pharmaceutical manufacturer in support of new drug applications for two important therapeutic products. Review scientists observed data inconsistencies and evidence of unacceptable laboratory practices in the study reports.

FDA's Bureau of Drugs requested a "for cause" inspection of the manufacturer's laboratories to determine the cause and extent of the discrepancies. A "for cause" inspection is one initiated at the request of an agency unit when there is reason to suspect a problem in an FDA-regulated product. The authority to make "for cause" inspections is a general one under the FFDCA, but one which had rarely been applied to animal laboratories.

Dr. Alexander M. Schmidt, Commissioner of Food and Drugs, in a statement in a Senate hearing on July 10, 1975, reported preliminary results of further agency investigations. (1) The findings indicated defects in design, conduct, and reporting of animal studies. "For cause" inspections were conducted at several laboratories and revealed similar problems. The nature and extent of the findings in these inspections raised questions about the validity of studies being submitted to the agency.

The deficiencies observed in these inspections were summarized in the preamble to the proposed Good Laboratory Practice Regulations (2) as follows:

1. Experiments were poorly conceived, carelessly executed, or inaccurately analyzed or reported....

2. Technical personnel were unaware of the importance of protocol adherence, accurate observations, accurate administration of test substance, and accurate recordkeeping and record transcription....

3. Management did not assure critical review of data or proper supervision of personnel....

4. Studies were impaired by protocol designs that did not allow the evaluation of all available data....

5. Assurance could not be given for the scientific qualifications and adequate training of personnel involved in the research study....

6. There was a disregard for the need to observe proper laboratory, animal care, and data management procedures....

7. Sponsors failed to monitor adequately the studies performed in whole or in part by contract testing laboratories....

8. Firms failed to verify the accuracy and completeness of scientific data in reports of nonclinical laboratory studies in a systematic manner before submission to the FDA.

The problems were so severe in Industrial Bio-Test Laboratories (IBT) and Biometric Testing Inc. that both laboratories ceased doing preclinical studies. IBT had been one of the largest testing laboratories in the United States, with thousands of its studies serving to support the safety of drugs, pesticides, and food additives. The FDA and the Environmental Protection Agency (EPA) began reviewing all the compounds that relied on IBT and Biometric Testing Inc. studies for support of safety. The agencies required the study sponsors to submit outside audits of the study data. From the audits of the IBT studies, EPA found 594 of 801 key studies, or 75%, invalid. (3) The FDA's Bureau of Foods found 24 of 66 IBT studies, or 36%, invalid. (4)

Criminal charges of fraud were brought against four IBT officials. Three of the officials were convicted; a mistrial was declared in the case of the fourth official because of illness. (5)

FDA'S RESPONSE TO THE PROBLEM

The conclusion that many studies on which the safety of regulated products had been based could be invalid was alarming to the FDA, to the EPA, to Congress, to the public, and to industry. Commissioner Schmidt established the Bioresearch Monitoring Program in early 1976 to develop a program that would deal with the problem of data validity not only in the area of safety studies but also in clinical testing. Congress voted a special appropriation of 16 million dollars and additional personnel to support the program.

A steering committee, chaired by the Associate Commissioner for Compliance and composed of the Associate Commissioners, the Bureau Directors, the Chief Counsel, the Director of the National Center for Toxicological Research, and the Executive Director for Regional Operations, directed the Bioresearch Monitoring Program. Four task forces, the Toxicology Laboratory Monitoring Task Force, the Investigator Sponsor Task Force, the Institutional Review Committee Task Force, and the Administrative Task Force handled different components of the program. The responsibility for developing a strategy to ensure the validity and reliability of all nonclinical laboratory studies to support the safety of FDA-regulated products was assigned to the Toxicology Monitoring Task Force. This task force was instructed to inventory all firms submitting research to the FDA and other involved federal agencies; to develop formal agreements with other agencies for the inspection of laboratories; to develop and publish standards for measuring the performance of research laboratories; to develop agency-wide enforcement strategies; and to develop plans for hiring, training, and assigning the new employees authorized by Congress for the program.

The Toxicology Monitoring Task Force chose the publication of Good Laboratory Practice (GLP) Regulations as the best approach for assuring study validity. Six other approaches were considered but were discounted as not feasible or efficient:

- One approach would have been to continue the program of "for cause" inspections. However, "for cause" inspections would be triggered only by perceived deficiencies in the data after submission to the agency and, thus, would not have provided systematic assurance that all studies were valid or guidance to laboratories on standards for conduct of studies.

- A second approach would have been to shift responsibility for nonclinical testing of regulated products to the FDA. Such a shift would have required congressional authorization, because the FFDCA clearly places this responsibility on the sponsor of the product. In addition, the costs of such a shift would have been prohibitive.

- The third approach considered was for the agency to publish detailed test protocols and procedures for studies on regulated products. However, this would have discouraged the use of informed scientific judgment in designing tests and inhibited the development of new toxicological methods.

- Another approach would have been to establish licensing procedures for testing laboratories, but developing uniform licensing criteria would have been very difficult, considering the variety of regulated products, of test types, and of laboratory facilities.

- Still another approach was the establishment of a full-time, on-site inspection program for laboratories similar to the U.S. Department of Agriculture's inspections of meat processing plants. Such a program was considered to be an inefficient use of the FDA's investigational resources, because many testing facilities are too small or too diversified to justify full-time, on-site monitoring.

· Consideration was also given to the publication of good laboratory practice guidelines, rather than regulations. While this would have provided the testing facilities with standards of conduct, it would not have given the agency an enforcement mechanism to ensure that the standards were met.

The regulations approach had several advantages. It was within the legal mandates of the agency and allowed efficient use of agency resources for ensuring compliance. It was also similar to the use of Good Manufacturing Practice (GMP) Regulations with which most of the regulated industries were already familiar. However, the main advantage was that the regulations approach focused on the process by which testing facilities carried out studies, rather than on the product being tested or the studies themselves. Thus, the use of scientific judgment in the planning and conduct of safety studies was not hampered, and the detail required for a focus on specific studies, or kinds of studies, was avoided.

Once the decision to establish GLP regulations had been made, a subcommittee was appointed to draft the regulations. This subcommittee was composed of individuals representing all the FDA bureaus and a variety of scientific disciplines. The subcommittee began its work with a rough draft that had already been prepared by personnel in the Bureau of Drugs. This early draft had used two independent, unsolicited sets of GLP guidelines submitted by G.D. Searle and Co. and the Pharmaceutical Manufacturers Association. The subcommittee's first draft was circulated to all FDA bureaus for comment, revised on the basis of these comments, and then circulated to other government agencies for comment. The subcommittee considered these comments in preparing the final draft which was published as the proposed GLP regulations on November 19, 1976. The proposed regulations were designated as a new part 3.e. of Chapter 21 of the Code of Federal Regulations, but the final regulations were codified as Part 58 (21 CFR Part 58).

FDA'S PROPOSED REGULATIONS

The purpose of the GLP regulations is to assure the quality and integrity of the data submitted to the FDA in support of the safety of regulated products. To this end, most of the requirements of the proposal would have been considered familiar and reasonable by any conscientious scientist. Protocols and standard operating procedures (SOPs), adequate facilities and equipment, full identification of test substances, proper animal care, equipment maintenance, accurate recording of observations, and accurate reporting of results are basic necessities for the conduct of a high-quality, valid toxicity, or any scientific study. The proposed regulations also placed a heavy emphasis on data recording and record and specimen retention to ensure that a study could be reconstructed at a later time if the need arose.

The proposed regulations went beyond these basic requirements for a valid study by requiring each study to have a study director, who would have "ultimate responsibility for implementation of the protocol and conduct of the study" [§ 3e.31(a)], and each testing facility to have a quality assurance unit to monitor conduct of studies. The concept of a quality assurance unit to monitor study conduct was a new one to most laboratories but a familiar one in manufacturing facilities operating under various Good Manufacturing Practice Regulations.

In addition, because the GLPs were regulations, the proposal identified the scope of the regulations, the authority under which they were promulgated, and the strategy for their enforcement.

Scope

The Toxicology Monitoring Task Force had not specified what types of studies would be considered to be within the scope of the GLPs. The subcommittee which drafted the regulations defined a nonclinical laboratory study as "any in vivo or in vitro experiment in which a test substance is studied prospectively in a test system under laboratory conditions to determine its safety" [§ 3e.3(d)]. The proposal explained that the term was to include only those studies conducted for submission to the FDA in support of an "application for a research or marketing permit." This latter term was a means of referring to the numerous categories of data required to be submitted to the agency, such as food and color additive petitions, new drug applications, and new animal drug applications. The studies covered by the regulations included all kinds of toxicity studies, from in vitro mutagenicity studies to acute, subchronic, and long-term toxicity/carcinogenicity studies and functionality/effectiveness studies when inadequate effectiveness might affect safety. Studies excluded from the scope of the regulations were those utilizing human subjects, clinical studies or field trials in animals, basic exploratory studies, or studies to determine physical or chemical properties of a test substance independent of a test system.

The proposal recognized that the scope might justifiably be defined on a different basis, possibly on a facilities basis, and asked for comments on whether specific types of testing facilities might be excluded from coverage by the regulations.

Enforcement Strategy

The basic mechanism of enforcement was to be inspection of testing facilities by FDA field investigators. The FDA's authority to conduct inspections of facilities engaged in interstate commerce of regulated products is well-established, and such inspections are the primary method of enforcement of the FFDCA. Under the proposal, studies performed by a testing facility which refused to permit inspection would not be accepted in support of an application for a research or marketing permit.

At the conclusion of an inspection, the FDA investigator notifies the facility of any deficiencies identified during the inspection in writing on Form 483, "Notice of Inspectional Observations," and in discussion with management. If the deficiencies were of a kind that might affect study validity, more formal warnings would be issued to the testing facility through a regulatory letter or a Notice of Adverse Findings.

Initial planning under the Bioresearch Monitoring Program called for each testing facility to be inspected yearly. It was later decided that a biennial inspection would suffice to ensure that all two-year studies would be inspected at least once while in progress.

When deficiencies were sufficient to affect the validity of a study, the proposal provided that the study would not be considered by the FDA in support of a research or marketing permit. However, the proposal noted that the data from such a study had to be submitted to the agency, and, that if it were adverse to the product, it might still be used as a basis for regulatory action. This difference in treatment was justified by the consideration that a bad study might reveal an adverse effect, but it could not establish absence of an adverse effect.

The final and most severe enforcement strategy under the proposal was the disqualification of a testing facility. Data from a disqualified facility would not be accepted in support of a research or marketing permit. The agency viewed this penalty as one that would only be employed in cases where the testing facility had severe, widespread deficiencies which raised questions about the validity of all the studies performed in the facility and where previous regulatory efforts had failed to bring the facility into compliance with the regulations. Unlike the other enforcement strategies, there was no specific authority for disqualification; the GLP regulations themselves established this authority.

Authority

The GLP regulations were issued under the general mandate of section 701(a) of the FFDCA, which empowers the Commissioner to promulgate regulations for the efficient enforcement of the Act. The Commissioner's power to issue regulations for determining that a clinical investigation of a drug intended for human use be scientifically reliable and valid [21 CFR 314.111(a)(5)] had been upheld by the Supreme Court in the decision *Weinberger vs. Hynson, Westcott and Dunning, Inc.*, 412 U.S. 609 (1973). The clinical investigations regulations had also been issued under section 701(a) of the FFDCA. It was further considered that the authority to issue GLP regulations gave the agency the authority to establish the terms on which it would accept nonclinical testing data; therefore, the proposed regulations provided for the rejection of studies if the testing facilities refused to permit inspection. The FDA already had the authority to compel inspection of nonclinical laboratories doing work on new drugs, new animal drugs, or medical devices. The FDA may inspect both manufacturing establishments and laboratories concerned with drugs and devices and examine research data on these products under section 704(a) of the FFDCA.

COMMENTS ON THE PROPOSAL AND THE FINAL REGULATIONS

More than 1000 individual items were contained in 22 oral responses from a two-day public hearing and 174 written responses to the proposal. Many responses commented on both general issues, such as scope, and specific details in individual sections and paragraphs. The preamble to the final regulations addressed these comments in detail, and modifications, both substantial and editorial, were included in the final regulations which were issued on December 22, 1978, and became effective June 20, 1979. (6)

Management and the Study Director

Comments on the responsibilities of the study director, as outlined in the proposal, identified many of these responsibilities as the prerogative of management. In response to these comments, a new section (§ 58.31) was included in the final regulations. This section established that the management of the testing facility had the responsibility for designating and replacing, if necessary, the study director; for providing a quality assurance unit and assuring that actions to correct deviations reported by the quality assurance unit are taken; for assuring that the personnel and the tools (e.g., facilities and equipment) are available as needed; and for assuring that test and control articles are appropriately identified.

Despite making management responsible for many areas that the proposal had assigned to the study director, the final regulations retained the concept of the study director as the single focus of responsibility for study conduct by redefining the function of the study director as: "overall responsibility for the technical conduct of the study, as well as for the interpretation, analysis, documentation and reporting of results, and represents the single point of study control" (§ 58.33).

The Quality Assurance Unit

Not surprisingly, many comments objected to the requirement for a quality assurance unit on the basis of increased costs, administrative burden, and interference with management prerogatives and informed scientific judgement of study directors. However, an alternative solution for study monitoring was not suggested.

The FDA retained the requirement for a quality assurance unit, or function, to monitor studies for conformance to the regulations. It was emphasized that the function was administrative, rather than scientific. The personnel responsible for quality assurance for a given study were required to be separate from, and independent of, the personnel responsible for the direction and conduct of that study.

Many commentators wanted the inspection records compiled by the quality assurance unit excluded from the records to be inspected by the agency on the basis that an inspection "might violate the constitutional privilege against compelled self-incrimination." The agency rejected this argument, because the privilege against compelled self-incrimination is not available to a collective entity, such as a business enterprise, or to an individual acting as a representative of a collective entity. The agency did, however, exclude the quality assurance unit's inspection records from inspection to encourage more forthrightness in the reports. The quality assurance unit was required to certify that the inspection of studies and final reports had been made by means of a signed statement to be included in the final report [§ 58.35(b)(7)].

Scope

In general, the comments on the proposed regulations generally sought limitations through exclusion of various classes of FDA-regulated products, such as medical devices; various types of facilities, such as academic and not-for-profit organizations; or various types of studies, such as short-term studies. These suggestions were rejected primarily because the basic purpose of the regulations, to ensure the validity of safety data submitted to the agency, would have been frustrated by excluding particular products, facilities, or studies from coverage. None of the commentators suggested an alternative overall approach to defining the scope of the regulations.

The scope adopted in the final regulations was only slightly changed from the proposal; the main difference was the exclusion of functionality studies from coverage.

Inspections

The major concerns of the commentators with respect to the actual inspection of facilities were the competence and scientific qualifications of the FDA investigators. In early inspections, both the "for cause" inspections prior to the proposal and the inspections made in the pilot program under the proposal, the agency assigned its most experienced field investigators and sent agency scientists to participate in the inspections. To further assure the competence of the investigators, a training program was established at the National Center for Toxicological Research for both field investigators and scientists. The compliance program for the GLPs also provides for scientific review in FDA headquarters of all GLP inspection reports.

That testing facilities still doubt the competence of some field investigators was evident in a comment on the 1987 revision of the GLPs, (7) which requested training in the GLPs for the FDA's field personnel.

Disqualification

Numerous comments were made on the provisions for disqualification of a testing facility (Subpart K). Although the proposal stated that the agency considered that it would only rarely invoke this penalty, it appeared from the objections that industry had interpreted these provisions to mean the agency would invoke disqualification frequently and for minor failures to comply with the regulations. On the basis of the objections, the purpose (§ 58.200) and the grounds for disqualification (§ 58.202) sections of Subpart K were extensively revised. The revision stated that the purposes of disqualification were:

1. to permit the exclusion of completed studies from consideration in safety evaluation until it could be shown that noncompliance with the regulations did not affect the validity of the study data, and

2. to permit the exclusion of studies completed after disqualification from consideration in safety evaluation until the facility could

demonstrate that it would conduct studies in compliance with the regulations.

Three grounds for disqualification were given in the final regulations; all three must be present to justify disqualification:

1. failure of the facility to comply with one or more of the GLP regulations or other regulations applying to facilities published in Chapter 21 of the Code of Federal Regulations,

2. the validity of the studies was adversely affected, and

3. failure to achieve compliance with regard to lesser regulatory actions, such as warnings or rejection of studies.

EVALUATION OF THE FDA PROGRAM

The proposed GLP regulations announced that the FDA would conduct a number of surveillance inspections of testing facilities, based on the requirements of the proposal, during November and December of 1976 and January of 1977. These inspections had the dual purpose of determining the status of the laboratories and evaluating the workability of the proposed regulations. The results of this pilot inspection program were analyzed and published by the FDA's Office of Planning and Evaluation. (8)

Forty-two laboratories were identified for inspection. Ongoing and completed studies would be examined as available. The inspections used a checklist which was divided into two parts, one part covering laboratory operations and the other study conduct. The checklist arbitrarily placed mixing and storing of test substances in the area of laboratory operations and distribution and characterization of the substances in study conduct.

In the completed survey, only 39 laboratories, with 67 studies, yielded usable data. Twenty-three of the testing facilities were sponsor laboratories, 11 contract laboratories, and 5 university laboratories. Forty-eight of the studies were completed and 19 ongoing. The findings showed that sponsor laboratories met 69% of the requirements, the contract laboratories met 56% of the requirements, and university laboratories met only 46% of the requirements.

Requirements in the areas of facilities, animal care, and personnel were the most often met, while fewest requirements were met in the areas of the quality assurance unit, mixing and storage of the test substances, and record retention.

Ongoing studies showed better adherence (73% of the requirements met) than did completed studies (57%). Animal care and test substance distribution showed the greatest degree of adherence. Low degrees of adherence were found in the quality assurance function and protocol-related requirements. The comments of the agency investigators indicated that testing facilities were already making changes in their ongoing studies to bring them into compliance.

Following publication of the final regulations, a second survey was conducted to measure compliance against the final requirements. (9) The study sample consisted of 17 sponsor laboratories, 10 contract laboratories and 1 university laboratory. The average compliance rate was 88%, with the deficiencies observed in sponsor and contract laboratories showing little difference. Compliance was measured both by the average compliance rate with the requirements of a section of the regulations or by the number of laboratories failing to meet one or more of the section's requirements. The following sections showed high compliance by both measurements: personnel, management, study director, general facilities, and facilities for animal care, handling test and control articles, laboratory operations, specimen and data storage, record retention, personnel and administration. Areas which showed low compliance by the same measures were: quality assurance units, maintenance and calibration of equipment, standard operating procedures, animal care (primarily the failure to analyze feed and water for interfering contaminants), test and control article characterization, mixtures of articles with carriers, study protocol, and study conduct (primarily failure to sign and date data sheets or to follow the protocol).

The results of these surveys indicated both the practicality of the regulations and the success of the vigorous efforts which most testing facilities were making to achieve compliance. The record of compliance continued to be good. The FDA reported, in its 1984 update of compliance results, (10) that 72% of the inspection reports since 1976 showed few or no substantial deviations from the regulations and 23% showed minor-to-significant deviations which could be corrected voluntarily by the testing facility. However, 4% of the reports showed significant deviations requiring corrective action within a specified period of time, and studies are still occasionally rejected because significant deviations render them invalid.

THE PROBLEM FROM EPA's PERSPECTIVE

The Environmental Protection Agency (EPA) had concerns similar to those of the FDA. Under Section 4 of the Toxic Substances Control Act (TSCA), the EPA evaluates laboratory data submitted to the agency regarding tests of the health effects of chemical substances and mixtures. The EPA also, under authority of the Federal Insecticide, Fungicide and Rodenticide Act (FIFRA), evaluates laboratory test data relating to hazards to humans arising from the use of a pesticide product when the agency evaluates pesticide registration applications.

The EPA was aware of the problems the FDA had uncovered in the mid-1970s relating to unacceptable laboratory practices. The EPA responded to the FDA's findings by forming the Toxicology Auditing Program in the agency's Office of Pesticide Programs. The EPA also held public hearings to solicit comments on how appropriate the agency's approach was to data quality assurance for pesticide testing.

In 1978, the EPA and FDA formalized both agencies' commitment to establish a coordinated quality assurance program through an interagency agreement. Under this agreement, the FDA provided assistance during EPA data audits. Between 1978 and 1979, the agencies performed 65 joint audits which indicated that some testing facilities did not follow good laboratory practices. The EPA referred some of these facilities to the Department of Justice for prosecution.

EPA'S PROPOSED REGULATIONS

Like the FDA, the EPA considered different approaches to assure that data submitted to the agency complied with necessary quality standards:

· Licensing or certification of laboratories was considered impractical for toxicology laboratories because of the great diversity and range of testing capabilities and the complex quality control procedures used in toxicology testing.

· A voluntary standard-setting scheme administered by the private sector was rejected because such schemes were considered practically unenforceable.

The EPA, as did the FDA, determined that the promulgation of Good Laboratory Practice Regulations would most effectively handle the problem of compliance with adequate control standards, and the agency published proposed health effects standards for testing under TSCA on May 9, 1979. (11) Proposed GLP regulations applicable to laboratory studies submitted to the EPA in compliance with FIFRA were published on April 18, 1980. (12) Supplemental GLP standards for the development of data on physical, chemical, persistence and ecological effects of chemical substances for which the EPA requires testing under section 4 of TSCA were published November 21, 1980. (13) The EPA took this action because the previously published GLPs for health effects testing did not address the analytical problems associated with physical, chemical, and persistence testing.

Differences Between EPA Proposed Regulations and FDA Regulations

When it issued proposed GLP regulations in 1978 and 1980, the EPA harmonized those regulations with the final GLP regulations which had been issued by the FDA in 1978. There were major differences, however, because the two agencies' approach to regulating laboratory studies differed. There were also differences in specific wording of various sections of the EPA's proposed regulations which differed from FDA's regulation because of the differing scope of the authority of each agency.

EPA's Final Regulations

The EPA's FIFRA and TSCA GLP regulations were both issued in final form on November 29, 1983. (14) The FIFRA GLP regulations were codified as 40 CFR 160, and the TSCA GLP regulations as 40 CFR 792. In terms of the TSCA GLPs, the final regulations incorporated the proposed GLPs issued on May 9, 1979 and November 21, 1980.

GLP REVISIONS IN THE 1980S

FDA Revisions

In 1984, the FDA proposed to revise its 1978 GLP regulations. The rationale for this revision was to clarify, amend, or delete provisions of the regulations in order to reduce the regulatory burden on testing facilities.

The FDA had found, during agency inspections, that most laboratories were complying with the GLP requirements -- indeed, that the violations it had noted in the mid 1970s were the exception, rather than the general rule -- and the agency thought that it could streamline the regulations without compromising the GLP program. The FDA had also received comments and questions about the GLP regulations which indicated that several GLP provisions did not significantly contribute to the quality and integrity of data submitted to the agency. At the same time, the agency was undertaking a review of its regulations to minimize regulatory burdens.

The FDA established a GLP Review Task Team to identify provisions in the regulations that could be amended or deleted, and this team recommended revisions to 36 GLP provisions. Recommendations were issued as a proposed rule on October 29, 1984. (15) The proposal made various changes to definitions to reduce the amount of paperwork required for nonclinical laboratory studies and to clarify earlier GLP provisions. Similar clarifications were made to the provisions delineating the definition and function of the study director and quality assurance unit.

In the 1984 proposal, changes were also made to information collection requirements subject to the Paperwork Reduction Act of 1980.

Modifications were made to the provisions regarding animal care, animal supply, and administrative and personnel facilities. Provisions regarding equipment design, maintenance and calibration of equipment, standard operating procedures, animal care, test and control article characterizations, and mixtures of articles with carriers were changed to allow more flexibility of laboratory operations. The section on laboratory protocols was amended to eliminate unnecessary protocol entries by allowing laboratories to identify the information in the protocol that is applicable to the articles being tested; the agency also deleted the requirement that selection of the test system be justified in the protocol. Other changes to the GLP regulations involved revisions to provisions regulating conduct of laboratory studies and the storage, retrieval, and retention of records.

The FDA received 33 comments on its proposed GLP revisions. After considering these comments, the agency issued its final GLP provisions on September 4, 1987. (16) Some of the comments received by the agency indicated a need to add new terms to the definition section of the regulations -- such as "study initiation" and "study completion" -- while others encouraged the FDA to retain the original GLP language in certain provisions rather than make the amendments the agency had proposed in 1984.

EPA Revisions

Among the comments received by the FDA, eight comments urged that agency to encourage the EPA to adopt similar revisions to its GLP regulations, which were now more stringent than the FDA's regulations. The FDA stated that the agency consulted with the EPA regarding the changes made to the FDA's regulations, and that the FDA would cooperate with the EPA when the latter agency revised its own GLP regulations. As a result of its own monitoring of GLP compliance, EPA agreed that its own GLP regulations could be streamlined without compromising the integrity of data submitted to the agency.

The EPA's proposed revisions to its FIFRA and TSCA GLP regulations were issued on December 28, 1987. (17) EPA agreed with the FDA that many GLP provisions could be amended to incorporate the changes that had been made by the FDA. In addition, the scope of the

FIFRA regulations was expanded to include environmental testing provisions that already existed in the TSCA GLPs, and to include product performance data (efficacy testing). The EPA also proposed to expand the scope of the TSCA GLP regulations to include field testing.

The EPA received 14 comments on its proposed GLP provisions, most of them supportive of the agency's proposed changes. Some changes were made to the proposed regulations in response to these comments, such as exempting from routine EPA inspections the records of quality assurance unit findings and problems, as well as records of corrective actions recommended and taken, except under special circumstances. The final versions of EPA's revisions to its GLPs were issued on August 17, 1989. (18)

EPA's proposed GLP revisions basically conformed to the changes the FDA had made in the latter agency's September 4, 1987 final rule. The major differences between the EPA proposals and the FDA GLPs continued to reflect the varying needs and responsibilities of each agency and the expanded scope of the EPA's regulations in light of the testing and test systems affected under the EPA's authority to require test data in support of research or marketing permits to include ecological effects, environmental and chemical fate, and efficacy testing in addition to health effects testing.

Other federal agencies, as well as international agencies and organizations, also developed GLP programs. The National Toxicology Program concluded that studies performed under contract to the program should be performed in compliance with GLPs and established a quality assurance function to monitor the laboratories and studies. In 1981, the Organization for Economic Cooperation and Development (OECD) developed GLP principles for studies performed for the European Economic Community countries. EC Council Directives between 1986 and 1988 adopted the OECD and required that all EC countries monitor and verify compliance with those standards.

In 1982, the Japanese Ministry of Health and Welfare issued GLP standards for safety studies on drugs; this was followed in 1984 by GLP standards issued for studies on industrial chemicals by the Japanese Ministry of International Trade and Industry and GLP standards for toxicological studies on agricultural chemicals by the Japanese Ministry of Agriculture, Forestry and Fisheries. There are differences in these regulations and guidelines which pose problems for sponsors planning studies to meet the requirements of different agencies or countries. (19)

As a solution to part of this problem, the FDA has developed Memoranda of Understanding (MOUs) with Canada (1979), Sweden (1979), Switzerland (1985), France (1986), Italy (1988), Germany (1988), the Netherlands (1988) and the United Kingdom (1988). These MOUs acknowledge mutual recognition of the adequacy of inspectional programs in the participating countries and permit exchange of data between the countries without need for independent verification by the recipient country.

REFERENCES

1. Statement by Alexander M. Schmidt, M.D., Commissioner, Food and Drug Administration, Public Health Service, Department of Health, Education and Welfare before the Subcommittee on Health, Committee on Labor and Public Welfare and Subcommittee on Administrative Practice and Procedure, Committee on the Judiciary, United States Senate, July 10, 1975.

2. Department of Health, Education and Welfare, Food and Drug Administration, Nonclinical Laboratory Studies: Proposed Regulations for Good Laboratory Practice. *Federal Register*, 41:51205-51230 (1976).

3. Environmental Protection Agency, Office of Pesticide Programs, Summary of the IBT Review Program. July 1983.

4. Survey by Dr. Jean M. Taylor.

5. IBT Data Falsification. *Food Chemistry News*, 25(24):2 (1983).

6. Department of Health, Education and Welfare, Food and Drug Administration, Good Laboratory Practice Regulations. *Federal Register*, 43:59985-60024 (1978).

7. Department of Health, Education and Welfare, Food and Drug Administration, Good Laboratory Practice Regulations. *Federal Register*, 52:33767-33782 (1987).

8. Food and Drug Administration, Office of Planning and Evaluation, OPE Study 42: Results of the Nonclinical Laboratory Good Laboratory Practices Pilot Compliance Program. September 1977.

9. Food and Drug Administration, Office of Planning and Evaluation, OPE Study 49: Results of the Toxicology Laboratory Inspection Program (January-March 1979). July 1979.

10. Department of Health, Education and Welfare, Food and Drug Administration, 21 CFR Part 58 Good Laboratory Practice Regulations; Proposed Rule. *Federal Register*, 49:43529-43537 (1984).

11. Environmental Protection Agency, Good Laboratory Practice Standards for Health Effects, Proposed Rule. *Federal Register*, 44:27362-27375 (1979).

12. Environmental Protection Agency, Pesticide Programs, Guidelines for Registering Pesticides in the United States; Proposed Good Laboratory Practice Guidelines for Toxicology Testing, Proposed Rule. *Federal Register*, 45:26373-26385 (1980).

13. Environmental Protection Agency, Physical, Chemical, Persistence, and Ecological Effects Testing; Good Laboratory Practice Standards, Proposed Rule. *Federal Register*, 45:77353-77365 (1980).

14. Environmental Protection Agency, Toxic Substances Control; Good Laboratory Practice Standards; Final Rule. *Federal Register*, 48:53921-53944 (1983). Environmental Protection Agency, Pesticides Program; Good Laboratory Practice Standards; Final Rule. *Federal Register*, 48:53945-53969 (1983).

15. Food and Drug Administration, Good Laboratory Practice Regulations, Proposed Rule. *Federal Register*, 49:43530-43537 (1984).

16. Food and Drug Administration, Good Laboratory Practice Regulations, Final Rule. *Federal Register*, 52:33768-33782 (1987).

17. Environmental Protection Agency, Federal Insecticide, Fungicide and Rodenticide Act (FIFRA); Good Laboratory Practice Standards, Proposed Rule. *Federal Register*, 52:48920-48933 (1987). Environmental Protection Agency, Toxic Substances Control Act (TSCA); Good Laboratory Practice Standards, Proposed Rule. *Federal Register*, 52:48933-48946 (1987).

18. Environmental Protection Agency, Toxic Substances Control Act (TSCA); Good Laboratory Practice Standards, Final Rule. *Federal Register*, 54:34034-34050 (1989). Environmental Protection Agency, Federal Insecticide, Fungicide and Rodenticide Act (FIFRA); Good Laboratory Practice Standards, Final Rule. *Federal Register*, 54:34052-34074 (1989).

19. For differences in worldwide GLP standards, see Chapter 3 of this book and also *International GLPs* by Robert S. DeWoskin and Stephanie M. Taulbee (Interpharm Press, 1993).

Chapter 2

FDA/GLP REGULATIONS

Wendell A. Peterson, J.D.

Parke-Davis Pharmaceutical Research Division of Warner-Lambert Company, Ann Arbor, Michigan

INTRODUCTION

Proposed Good Laboratory Practice (GLP) regulations were published in 1976 (1). Final regulations were published in 1978 (2). The regulations were revised in 1980 (3), in 1987 (4), twice in 1989 (5, 6), and in 1991 (7).

GLP regulations (8) are promulgated by the Commissioner of the U.S. Food and Drug Administration (FDA) under general authority granted by Section 701(a) of the Federal Food, Drug, and Cosmetic (FD&C) Act (9). Unlike Current Good Manufacturing Practice regulations (10), which can be referenced back to specific statutory language (the words "current good manufacturing practice" in Section 501(a)(2)(B)), the phrase "good laboratory practice" does not appear in the FD&C Act. Rather, the GLP regulations are issued under the FDA Commissioner's implied powers to prescribe standards for the conduct of studies designed to establish the safety of products regulated by the FDA.

This chapter provides a general discussion of all aspects of FDA's GLP regulations, as amended to September 13, 1991. Where appropriate, FDA interpretations are presented for specific sections of the GLP regulations. For critical parts of the regulations, a more in-depth discussion is provided, including means for implementation and an evaluation of positive and negative impacts on the conduct of GLP-regulated studies.

SUBPART A -- GENERAL PROVISIONS

§ 58.1 Scope

(a) This part prescribes good laboratory practices for conducting nonclinical laboratory studies that support or are intended to support applications for research or marketing permits for products regulated by the Food and Drug Administration, including food and color additives, animal food additives, human and animal drugs, medical devices for human use, biological products, and electronic products. Compliance with this part is intended to assure the quality and integrity of the safety data filed pursuant to sections 406, 408, 409, 502, 503, 505, 506, 507, 510, 512-516, 518-520, 706, and 801 of the Federal Food, Drug, and Cosmetic Act and sections 351 and 354-360F of the Public Health Service Act.

(b) References in this part to regulatory sections of the Code of Federal Regulations are to Chapter I of Title 21, unless otherwise noted.

The preamble to the GLP regulations (2), the report of a series of three briefing sessions on the GLP regulations which were conducted by FDA on May 1, 2, and 3, 1979 [11], and the collection of responses by Dr. Paul Lepore (FDA spokesman on GLP issues) to questions about the GLP regulations (12), have defined the types of studies to which the GLP regulations apply. In general, *all* of the following conditions must exist before a study will be regulated by GLP:

1. study of a product regulated by the FDA (except cosmetics).
2. in vivo or in vitro study.

3. study in which the FDA-regulated product is administered or added to nonhuman animals, plants, microorganisms, or subparts of the preceding.
4. study results submitted or intended to be submitted to FDA in support of (i.e., as the basis for) the approval of an application for a research or marketing permit.
5. study results may be used to predict adverse effects of and/or to establish safe use characteristics for the FDA-regulated product.

The FDA has made clear that the duration of the study and the place where the study is conducted do not determine whether the study is GLP-regulated. Thus, the GLPs apply to short-term studies (e.g., median lethal dose studies and irritation studies) as well as long-term studies which meet all of the above criteria, and the GLPs apply to such studies whether conducted in a manufacturer's laboratories, in a university laboratory, or at a contract or subcontract facility. FDA also expects GLP compliance for studies conducted in foreign countries as well as for those conducted within the United States.

Without attempting to provide a comprehensive listing, the following are examples of studies to which the GLPs can apply:

Ames Test
E. coli mutagenicity
Sister chromatid exchange
Bone marrow cytogenetic
In vivo cytogenetic
In vitro mutation
In vivo micronucleus
Chromosomal aberration
Median lethal dose (LD_{50})
Acute dermal toxicity
28-Day dermal toxicity
Dermal irritation
Eye irritation
Venous irritation
Muscle irritation

Intraarterial tolerance

Guinea pig maximization

Phototoxicity

Ototoxicity

Dependency tests on known or suspected addictive drugs

Target animal absorption, distribution, metabolism and excretion (ADME)

Subchronic (up to 13 weeks' duration, multiple dosing, any route of administration)

Chronic (6 months or longer in duration, multiple dosing, any route of administration)

Study of fertility in early embryonic development (previously referenced as Segment I)

Study of embryo-fetal development (formerly referred to as Segment II)

Perinatal/postnatal (formerly referred to as Segment III)

To reiterate, the foregoing list is only intended to illustrate the wide range of studies which may be GLP-regulated.

Examples of studies that are not within the scope of the GLP regulations include:

Pharmacology experiments

Basic research

Dose-range finding studies

Studies to develop new methodologies

Human or animal efficacy studies

Chemical assays for quality control of commercial products

Stability tests on finished dosage forms and products

Tests for conformance to pharmacopeial standards

Exploratory studies on viruses and cell biology

Tests of functionality and/or appropriateness of food additives

Tests of extractability of polymeric materials that contact food

Chemical tests used to derive the specifications of marketed food products

Studies on medical devices that do not come in contact with or are not implanted in humans

Tests of diagnostic products
Chemical and physical tests of radiation products
Tests conducted for the release of licensed biological products

The foregoing list is also intended to be illustrative, and not comprehensive.

A facility which conducts both GLP-regulated and non-GLP-regulated studies should think carefully about attempting to maintain a dual standard in any one laboratory or with any one group of laboratory workers. In the author's experience, such a dual standard is very difficult to maintain without carryover of non-GLP standards to GLP-regulated work. In such a case it may be far better to maintain a general GLP standard (e.g., data collection, recordkeeping, etc.) for all work in the laboratory but, perhaps, allow exceptions for the non-GLP studies in areas such as quality assurance (QA) inspections and analytical requirements for test and control articles and article/carrier mixtures.

The effective date of the GLP regulations was June 20, 1979. The regulations did not apply retroactively. Therefore, studies begun and completed prior to the effective date were not required to comply with the GLPs even if submitted to FDA on or after June 20, 1979. For studies in progress on June 20, 1979, only those portions of the study done on or after June 20 were required to be done in compliance with the regulations. Of course, those studies initiated on or after the effective date were to be done in full compliance with the GLPs.

§ 58.3 Definitions

A good understanding of the definitions in section 58.3 is critical to an interpretation of many of the other sections of the regulations.

An illustration of the importance of a definition to regulatory interpretation can be found in Environmental Protection Agency (EPA) regulations issued under the Resource Conservation and Recovery Act (RCRA). Among other things, these regulations are designed to regulate the disposal of "solid waste." Anyone relying on the normal definition of "solid" in interpreting RCRA requirements would make a grave error because "solid" is defined in RCRA to include solid, liquid, semisolid, and contained gaseous materials.

Although there is nothing in the definitions section of the GLP regulations to rival RCRA's rewriting of the basic laws of chemistry and physics, a clear understanding of GLP definitions is essential to a proper interpretation of GLP requirements.

As used in this part, the following terms shall have the meanings specified:

(a) "Act" means the Federal Food, Drug, and Cosmetic Act, as amended (secs. 201-902, 52 Stat. 1040 et seq., as amended (21 U.S.C. 321-392)).

(b) "Test Article" means any food additive, color additive, drug, biological product, electronic product, medical device for human use, or any other article subject to regulation under the act or under sections 351 and 354-360F of the Public Health Service Act.

Note the wide range of FDA-regulated products to which the GLPs apply.

(c) "Control article" means any food additive, color additive, drug, biological product, electronic product, medical device for human use, or any article other than a test article, feed, or water that is administered to the test system in the course of a nonclinical laboratory study for the purpose of establishing a basis for comparison with the test article.

The term "control article" refers to materials that are administered or added to the typical "control group" which is part of most safety studies. The term includes materials commonly referred to as "positive controls" (e.g., a marketed drug which is administered or added to a positive control group as part of a study of an investigational drug of the same therapeutic category) as well as vehicles, solvents, and other carrier materials (other than water and animal diets) when such materials are given to control groups.

Although the GLP revisions of 1987 excluded animal feed and water from the definition of control article, it would appear that such common vehicles as saline solutions and carboxymethylcellulose solutions still fall within the definition. Such a strict definition of the term for such innocuous vehicles as saline solutions is quite burdensome when one considers the requirements for control articles which are found in other sections of the GLPs: characterization (§ 58.105(a)), stability testing (§ 58.105(b)), sample retention (§ 58.105(d)) and inventory (§ 58.107(d)). It does not appear that this comprehensive definition is enforced by FDA field investigators in the course of GLP inspections.

Positive controls (usually known mutagens) used in mutagenicity studies also fall outside the definition of control article because they are administered to control groups for the purpose of establishing the ability of the assay to detect mutagenic activity and not for the purpose of "establishing a basis for comparison with the test article."

(d) "Nonclinical laboratory study" means in vivo or in vitro experiments in which test articles are studied prospectively in test systems under laboratory conditions to determine their safety. The term does not include studies utilizing human subjects or clinical studies or field trials in animals. The term does not include basic exploratory studies carried out to determine whether a test article has any potential utility or to determine physical or chemical characteristics of a test article.

Many of the issues relating to the definition of "nonclinical laboratory study" were addressed in the discussion of GLP § 58.1 (Scope). "Field trials in animals" includes all efficacy studies of new animal drugs. Such studies are outside the scope of the GLP regulations. This is consistent with the GLP exemption for human clinical trials. The exemption for "basic exploratory studies carried out to determine whether a test article has any potential utility" would extend to early screening studies of a test article, the results of which are used to determine whether a test article merits further development.

GLP § 58.105(a) requires that all test articles be appropriately characterized. Compliance requires documentation that characterization has been done. However, the tests conducted to provide this documentation are not GLP-regulated, although such tests will in many instances be subject to CGMP standards (e.g., when the test article will also be used in human clinical studies).

The GLP revisions of 1987 modified slightly the definition of "nonclinical laboratory study" by changing a few nouns, verbs, and adjectives from singular to plural form. This now permits the conduct of several experiments using the same test article under a single, comprehensive protocol or the concurrent test of several test articles using a single common procedure under a single protocol.

(e)   "Application for research or marketing permit" includes:

(1)   A color additive petition, described in part 71.

(2)   A food additive petition, described in parts 171 and 571.

(3)   Data and information regarding a substance submitted as part of the procedures for establishing that a substance is generally recognized as safe for use, which use results or may reasonably be expected to result, directly or indirectly, in its becoming a component or otherwise affecting the characteristics of any food, described in §§ 170.35 and 570.35.

(4)   Data and information regarding a food additive submitted as part of the procedures regarding food additives permitted to be used on an interim basis pending additional study, described in § 180.1.

(5)   An "investigational new drug application," described in Part 312 of this chapter.

(6)   A "new drug application," described in part 314.

(7)   Data and information regarding an over-the-counter drug for human use, submitted as part of the procedures for classifying such drugs as generally recognized as safe and effective and not misbranded, described in part 330.

(8)   Data and information about a substance submitted as part of the procedures for establishing a tolerance for unavoidable contaminants in food and food-packaging materials, described in parts 109 and 509.

(9)   Data and information regarding an antibiotic drug submitted as part of the procedures for issuing, amending, or repealing

regulations for such drugs, described in § 314.300 of this chapter.

(10) A "Notice of Claimed Investigational Exemption for a New Animal Drug," described in part 511.

(11) A "new animal drug application," described in part 514.

(12) (Reserved)

(13) An "application for a biological product license," described in part 601.

(14) An "application for an investigational device exemption," described in part 812.

(15) An "Application for Premarket Approval of a Medical Device," described in section 515 of the act.

(16) A "Product Development Protocol for a Medical Device," described in section 515 of the act.

(17) Data and information regarding a medical device submitted as part of the procedures for classifying such devices, described in part 860.

(18) Data and information regarding a medical device submitted as part of the procedures for establishing, amending, or repealing a performance standard for such devices, described in part 861.

(19) Data and information regarding an electronic product submitted as part of the procedures for obtaining an exemption from notification of a radiation safety defect or failure of compliance with a radiation safety performance standard, described in subpart D of part 1003.

(20) Data and information regarding an electronic product submitted as part of the procedures for establishing, amending, or repealing a standard for such product, described in section 358 of the Public Health Service Act.

(21) Data and information regarding an electronic product submitted as part of the procedures for obtaining a variance from any electronic product performance standard as described in § 1010.4.

(22) Data and information regarding an electronic product submitted as part of the procedures for granting, amending,

or extending an exemption from any electronic product performance standard, as described in § 1010.5.

This section of the GLPs describes the various types of submissions to FDA which include safety information derived from studies which must be conducted in accordance with the GLP regulations.

  (f) "Sponsor" means:
  (1) A person who initiates and supports, by provision of financial or other resources, a nonclinical laboratory study;
  (2) A person who submits a nonclinical study to the Food and Drug Administration in support of an application for a research or marketing permit; or
  (3) A testing facility, if it both initiates and actually conducts the study.

The definition of "sponsor" indicates who bears ultimate responsibility for a nonclinical laboratory study. A sponsor may assign the job of actual study conduct and/or reporting, but ultimate responsibility for the study cannot be delegated. Thus, the sponsor must assure that a nonclinical laboratory study is conducted in compliance with GLP standards, and the sponsor must supply the statement of GLP compliance or description of GLP non-compliance (conforming amendments statement) which must accompany the submission to FDA of the results of a nonclinical laboratory study (Section XI). The definition does not preclude joint sponsorship of a study.

  (g) "Testing facility" means a person who actually conducts a nonclinical laboratory study, i.e., actually uses the test article in a test system. "Testing facility" includes any establishment required to register under section 510 of the act that conducts nonclinical laboratory studies and any consulting laboratory described in section 704 of the act that conducts such studies. "Testing facility" encompasses only those operational units that are being or have been used to conduct nonclinical laboratory studies.

If a facility conducts nonclinical laboratory studies, it is a "testing facility" and is subject to inspection by FDA to determine its GLP compliance status. If a facility conducts nonclinical laboratory studies as well as studies which do not meet the definition of nonclinical laboratory study, then only those portions of the facility that conduct nonclinical laboratory studies are subject to a GLP inspection by FDA. The portions of the facility that conduct other than nonclinical laboratory studies are not subject to inspection by FDA unless FDA has inspectional authority under some other set of regulations.

> (h) "Person" includes an individual, partnership, corporation, association, scientific or academic establishment, government agency, or organizational unit thereof, and any other legal entity.

This all-encompassing definition of "person" precludes the exemption of any person or legal entity from the definition of "sponsor" or "testing facility" if that person or other legal entity meets the definitions of those two terms.

> (i) "Test system" means any animal, plant, microorganism, or subparts thereof to which the test or control article is administered or added for study. "Test system" also includes appropriate groups or components of the system not treated with the test or control articles.

In most instances the "test system" will be self-evident (e.g., the animal to which the test article is administered or applied). However, studies with microorganisms sometimes present difficulty in defining the test system. In the case of the Ames test, for example, the test system is not merely the colonies of salmonella or yeast, but includes in addition the culture medium, metabolic activation agent (if any), biotin, histidine, and buffer (if any). The last sentence of the definition makes it clear that untreated control groups also meet the definition of test system even though a test or control article is not administered or applied to such groups.

(j)   "Specimen" means any material derived from a test system for examination or analysis.

In most instances the "specimens" will be self-evident (e.g., samples of blood, plasma, serum, urine, spinal fluid, aqueous humor, organs, tissues, and tissue fractions which are taken from a test system with the intention of performing an examination or analysis). In other instances the definition may not be as clear. For example, the assay plates used in the mammalian cell transformation assay and the mammalian point mutation assay are considered specimens even though they bear many of the attributes of a test system. For these assays, the originally plated cells plus media and excipients are the test system. However, after treatment with the test or control article the plates are stained and transformed cells are enumerated. The plates then become "material derived from the test system for examination or analysis," in other words, specimens.

Care should be taken to distinguish specimen from "raw data," since GLP requirements differ for each. For example, it is often erroneously stated that a microscopic slide is raw data when, in fact, it is a specimen.

(k)   "Raw data" means any laboratory worksheets, records, memoranda, notes, or exact copies thereof, that are the result of original observations and activities of a nonclinical laboratory study and are necessary for the reconstruction and evaluation of the report of that study. In the event that exact transcripts of raw data have been prepared (e.g., tapes which have been transcribed verbatim, dated, and verified accurate by signature), the exact copy or exact transcript may be substituted for the original source as raw data. "Raw data" may include photographs, microfilm or microfiche copies, computer printouts, magnetic media, including dictated observations, and recorded data from automated instruments.

Examples of raw data include records of animal receipt, records of animal quarantine, results of environmental monitoring, instrument calibration records, original recordings of such parameters as animal body weights or food consumption values, handwritten transcriptions to paper records of information displayed as a digital read-out on automated equipment, high-performance liquid chromatography (HPLC) tracings, integrator output from HPLC equipment, recorded clinical observations, a photograph of a lesion noted at autopsy, a pathologist's written or tape-recorded diagnosis of a microscopic slide, printed paper tapes containing values generated by hematology and blood chemistry equipment, and electrocardiographic tracings. These are only examples; the reader could expand the list ten- or one hundred-fold.

Microfilm and microfiche copies, carbon copies, or photocopies of original raw data may be substituted for the original raw data so long as they are exact and legible copies.

Cage cards which contain information such as animal number, study number, treatment group, etc. are not raw data as long as no original observations are recorded on the card. Nor are transformations of raw data (e.g., calculations of mean and standard deviation or other statistical values) considered raw data because they can always be recalculated from the original raw data.

In the case of handwritten raw data, the original recording of information on paper constitutes the raw data which must be retained under § 58.190(a) of the regulations. Any subsequent transcriptions of this information will not substitute for the originally recorded information. Scientists and technicians will sometimes record raw data on scraps of paper or even on paper towels. Their intention is to neatly transcribe the information to official data forms at a later time and to discard the originally recorded data. This practice is to be discouraged, because the scraps of paper or paper towels are the real raw data and must be retained.

FDA has indicated that a pathologist's interim microscopic diagnoses are not raw data because such diagnoses are not "necessary for the reconstruction and evaluation of the report of (a) study." Only when the pathologist signs off on a final diagnosis does that diagnosis become raw data.

The provision in the definition of raw data for the substitution of exact transcripts of raw data for the original has been narrowly construed by FDA. It applies only to the verbatim transcription of tape-recorded information (e.g., a pathologist's voice recording of a microscopic diagnosis or veterinarian's voice recording of a clinical observation) which is dated and verified as accurate by signature. In this case the original tape recording need not be retained.

If raw data is transcribed to a computer data base, neither the electronically stored data nor its paper printout can substitute for the original. However, information entered into the computer by direct data capture offers two options. The laboratory may elect to treat the electronically recorded information or a hard copy printout of the information as raw data. If hard copy is retained, the magnetic media can be discarded or reused. If a laboratory elects to treat the magnetic media as raw data, it must retain an ability to display that data in readable form for the entire period during which that information is required to be retained. (See § 58.195 for a definition of required retention periods.) If a change in computer systems would entail the loss of the ability to display electronically stored data, then the laboratory should generate hard copies of the data before the computer systems are changed.

> (l)   "Quality assurance unit" means any person or organizational element, except the study director, designated by testing facility management to perform the duties relating to quality assurance of nonclinical laboratory studies.

Note the language "any person...except the study director." When read in conjunction with GLP § 58.35(a), it is clear that the person or persons designated to perform quality assurance functions need not be full-time quality assurance personnel. This flexibility is provided primarily to accommodate smaller laboratories where the volume of GLP-regulated work is not sufficient to justify a full-time quality assurance person. A person from the pharmacology department, for example, can perform the quality assurance function for toxicology studies on a part-time basis, but spend the rest of his or her time in the conduct of pharmacology studies. Where the volume of work is sufficient to justify employing one or more full-time quality assurance professionals, that is the preferred arrangement. Such an arrangement provides the degree of independence so important to the success of any quality program, removes the possibility that the demands of the part-time quality assurance person's other responsibilities will interfere with his or her performance of the quality assurance function, and allows more time for the development of expert audit and inspection skills. The author is aware of no major testing facility in the United States where the quality assurance unit is routinely staffed by persons other than full-time quality assurance professionals, although there are instances where the full-time quality assurance staff is supplemented by temporary assignments from other departments.

Issues relating to the "quality assurance unit" will be addressed in greater depth in the later discussion of GLP § 58.35.

> (m) "Study director" means the individual responsible for the overall conduct of a nonclinical laboratory study.

Note the words "the individual." There may be only one designated study director for any one study at any one time. It is not permissible, for example, to appoint a "study director" and an "assistant study director," but it is permissible to name an "alternate study director" who will serve as study director only in the absence of the "study director." It is FDA's intent that the "study director" serve as the single point of study control.

For a detailed description of the "study director" role, see the later discussion for GLP § 58.33.

> (n) "Batch" means a specific quantity or lot of a test or control article that has been characterized according to § 58.105(a).

The GLP definition of "batch" differs from that found in FDA's Current Good Manufacturing Practice (cGMP) regulations (§ 210.3(b)(2)) (10). The cGMP definition relates any one "batch" to a defined cycle of manufacture. The GLP definition, on the other hand, relates "batch" to a characterization process. Thus, for example, a GLP "batch" may be part of a cGMP "batch" or may be the result of a combination of two or more cGMP "batches." The only GLP requirement is that a "batch" be characterized as to identity, strength, purity, and composition or other appropriate characteristics.

> (o) "Study initiation date" means the date the protocol is signed by the study director.

(p) "Study completion date" means the date the final report is
signed by the study director.

§ 58.10 Applicability to Studies Performed Under Grants and Contracts

When a sponsor conducting a nonclinical laboratory study
intended to be submitted to or reviewed by the Food and Drug
Administration utilizes the services of a consulting laboratory,
contractor, or grantee to perform an analysis or other service, it
shall notify the consulting laboratory, contractor, or grantee that
the service is part of a nonclinical laboratory study that must be
conducted in compliance with the provisions of this part.

The notification required by this section should be in writing. The
form of the writing is not important from a GLP standpoint, but it may
be advantageous to put the notification into a legally binding document
(e.g., contract). Alternatively, the notification may appear, for example,
in a study protocol signed by the sponsor or in a letter from the sponsor
to the contractor.

High-volume contract laboratories often perform both GLP-regulated
and non-GLP-regulated studies, so it is important to specify if a study is
to be conducted under GLP conditions.

Some contract laboratories and professional consultants (e.g.,
veterinary ophthalmologists and pathologists), may not be familiar with
the GLP regulations. In such cases, mere notification of a requirement
to provide GLP-complying services may not be sufficient. It is advisable
to spend time with contractors and professional consultants to review in
detail the GLP requirements that will apply to the work they will
perform.   It is especially important to review with them the GLP
requirements for documentation and document retention.

§ 58.15 Inspection of a Testing Facility

(a) A testing facility shall permit an authorized employee of the
Food and Drug Administration, at reasonable times and in a
reasonable manner, to inspect the facility and to inspect (and
in the case of records also to copy) all records and specimens

required to be maintained regarding studies within the scope of this part. The records inspection and copying requirements shall not apply to quality assurance unit records of findings and problems, or to actions recommended and taken.

(b) The Food and Drug Administration will not consider a nonclinical laboratory study in support of an application for a research or marketing permit if the testing facility refuses to permit inspection. The determination that a nonclinical laboratory study will not be considered in support of an application for a research or marketing permit does not, however, relieve the applicant for such a permit of any obligation under any applicable statute or regulation to submit the results of the study to the Food and Drug Administration.

All laboratories operating within the United States which conduct nonclinical laboratory studies are subject to inspection by FDA. Such inspections may include an inspection of laboratory facilities, laboratory records, and specimens. FDA, however, has no legal authority to conduct such inspections outside of the United States. FDA inspections of laboratories outside the United States do occur, but only after a request from the FDA to conduct such an inspection has received consent of the laboratory involved. When a sponsor uses the services of a contract laboratory, consulting laboratory, contractor, or grantee to conduct all or any portion of a nonclinical laboratory study, it is advisable to obtain written consent of such groups to submit to inspection by the FDA on request, as a condition of placing work with the contractor or grantee. This is especially true in the case of contractors or grantees who do not routinely conduct nonclinical laboratory studies, and may be unaware of their obligation to permit such inspection, and may not be inclined to consent to inspection.

If a testing facility refuses to permit an FDA inspection, none of the nonclinical laboratory studies, or parts of studies, conducted by that laboratory will be considered in support of an application for a research or marketing permit. The results of such studies must be submitted to

FDA, but the results would not be accepted as evidence of the safety of the test article. However, such results could be used by FDA to support a finding that the test article was not safe.

FDA inspections must occur at "reasonable times," which is generally defined as during normal business hours. Inspections must also be conducted in a "reasonable manner" which would include adherence by the inspector to all laboratory safety policies (e.g., wearing of safety goggles) and compliance with normal requirements for donning protective apparel (gown or lab coat, hat, mask, shoe covers, etc.) before entering animal housing areas.

FDA inspectional authority includes the right to copy records and to collect samples. It is discretionary with the inspected laboratory whether to charge for copies of records or to provide them to the inspector free-of-charge. The same is true with regard to the inspector's request for samples, although requests for samples are rare during GLP inspections. In most cases, laboratories will provide samples and copies of documents free-of-charge unless FDA requests for these are excessive.

Quality assurance (QA) unit records are exempt from routine FDA inspection and copying authority on the theory that such records are more likely to be complete and candid if they are exempt from review by the FDA. This exemption extends only to records of QA inspection and audit findings and records of corrective actions recommended and taken. All other QA records are subject to inspection and copying by FDA.

The one exception to FDA's policy of not seeking access to QA records of findings and problems or of corrective actions recommended and taken is that FDA may seek production of these reports in litigation under applicable procedural rules. Therefore, the quality assurance unit should seek the advice of house counsel as to the retention period for such records.

Before 1992 FDA normally provided at least one week advance notice of a GLP inspection except in the case of "for cause" inspections which usually occurred without advance notice. However, current FDA policy is to conduct all GLP inspections in the United States without advance notice. Generally FDA continues to provide advance notice of GLP inspections outside the United States.

For an excellent discussion of legal issues surrounding FDA's inspectional authority, see Volumes 16 and 52 of this series (13, 14).

SUBPART B -- ORGANIZATION AND PERSONNEL

§ 58.29 Personnel

    (a)   Each individual engaged in the conduct of or responsible for the supervision of a nonclinical laboratory study shall have education, training, and experience, or combination thereof, to enable that individual to perform the assigned functions.

    (b)   Each testing facility shall maintain a current summary of training and experience and job description for each individual engaged in or supervising the conduct of a nonclinical laboratory study.

FDA has refrained from specifying exactly what scientific disciplines, education, training, or expertise qualify individuals to participate in the conduct of a nonclinical laboratory study. These factors vary from study to study, and FDA has merely indicated that the question of employee qualifications should be carefully considered by laboratory management. Laboratory management, therefore, has considerable latitude to define job qualifications. Any reputable laboratory will find it to be in its own best interest to hire competent individuals and to provide adequate on-the-job training to qualify those individuals to perform their assigned duties. FDA is not likely to make an issue of employee qualifications unless an inspection reveals an obvious case of employee incompetence.

Documentation of employee qualifications should include, at a minimum, an educational history for each employee, an employee's employment history to the extent that prior employment has a bearing on competence to perform the employee's current job assignment, and a description of any additional on-the-job training provided to the employee. Any format is acceptable for documentation of employee qualifications as long as all relevant information is included. The degree of detail associated with documentation of on-the-job training varies widely from laboratory to laboratory. Some laboratories document supervisor/trainer sign-off for completion of training in each element of an employee's current job description. Other laboratories merely document successful completion of an employee's initial probationary period. Documentation should be updated periodically to reflect changes in educational background and any additional training provided to the employee.

(c)   There shall be a sufficient number of personnel for the timely and proper conduct of the study according to the protocol.

The requirement for adequate numbers of personnel was included in the GLP regulations as a result of FDA's pre-GLP inspection of a laboratory that had taken on more work than its employees could properly perform.   The result, according to FDA, was poor quality or, even, fraudulent data.

In FDA's opinion, a shortage of qualified personnel can lead to inadequate or incomplete monitoring of a study and to delayed preparation and analysis of study results.   The numbers of personnel conducting a study should be sufficient to avoid such problems.

Today it is unlikely that a laboratory would be prospectively cited by FDA for inadequate numbers of personnel. Any citation in this area is more likely to be retrospective and based on actual evidence of poor quality work related to inadequate numbers of personnel.

(d)   Personnel shall take necessary personal sanitation and health precautions designed to avoid contamination of test and control articles and test systems.

(e)   Personnel engaged in a nonclinical laboratory study shall wear clothing appropriate for the duties they perform. Such clothing shall be changed as often as necessary to prevent microbiological, radiological, or chemical contamination of test systems and test and control articles.

Although these sections of the GLPs are designed to protect test and control articles and test systems, laboratory management should also take into account federal and state requirements for protection of the health and safety of employees. The minimum acceptable protective apparel for employees working with test and control articles and with animals is a laboratory coat over street clothing. Many laboratories provide uniforms. A sufficient supply of clean apparel should be provided by the company to allow frequent changes if suggested by the hazards of the materials, or if necessary to protect against cross-contamination. The wearing of hats, gloves, masks, and shoe covers (preferably of the disposable variety) is highly recommended. Enough of these items should be provided to permit changes when moving between rooms. Safety glasses or protective goggles will be appropriate for some hazardous operations.

A laboratory should have a generic policy for the safe handling of chemicals plus special policies for work with hazardous materials.

Refer to NIH Publication No. 85-23, *Guide for the Care and Use of Laboratory Animals*; NIH Publication No. 81-2385, *NIH Guidelines for the Laboratory Use of Chemical Carcinogens*; and the Public Health Service *Biosafety Guidelines for Microbiological and Biomedical Laboratories* for additional discussion of these issues.

(f)   Any individual found at any time to have an illness that may adversely affect the quality and integrity of the nonclinical laboratory study shall be excluded from direct contact with test systems, test and control articles and any other operation or function that may adversely affect the study until the condition is corrected. All personnel shall be instructed to report to their immediate supervisors any health or medical conditions that may reasonably be considered to have an adverse effect on a nonclinical laboratory study.

The potential for spread of disease organisms from animals to humans and vice versa is not obvious to most people. These so-called zoonotic diseases include agents of all the major categories of infectious organisms: viruses, bacteria, parasites, and fungi. Infectious hazards are insidious and, therefore, safe practices should be habitual and strictly enforced. All employees should be instructed as to the nature of these hazards and the means to take to protect animals and themselves from infection. Employees should also be instructed to report all personal illnesses to their supervisor. The supervisor can then determine whether it would be appropriate for the employee to have contact with test and control articles and test systems.

§ 58.31 Testing Facility Management

For each nonclinical laboratory study, testing facility management shall:

(a) Designate a study director as described in § 58.33, before the study is initiated.

(b) Replace the study director promptly if it becomes necessary to do so during the conduct of a study.

A study director for each study can be designated in the study protocol which is approved by management or in separate documentation which is signed by management. As mentioned in the discussion of GLP definitions, only one person may be designated as study director. Study co-directors are not permissible, but an alternate study director may be designated. If the study director must be replaced, this may be accomplished by protocol amendment (if the original study director was designated in the protocol) or by separate documentation (if separate documentation was used to appoint the original study director).

(c) Assure that there is a quality assurance unit as described in § 58.35.

(d) Assure that test and control articles or mixtures have been appropriately tested for identity, strength, purity, stability, and uniformity, as applicable.

(e) Assure that personnel, resources, facilities, equipment, materials, and methodologies are available as scheduled.

(f) Assure that personnel clearly understand the functions they are to perform.

(g) Assure that any deviations from these regulations reported by the quality assurance unit are communicated to the study director and corrective actions are taken and documented.

These duties, which are more administrative than scientific, are the responsibility of management. "Management" will generally be defined as the person or persons who have authority within an organization to effect whatever changes are necessary to assure that these duties are adequately discharged. Identification of such persons will vary depending on the structure of each organization. Management may, of course, delegate these duties to others within the organization. Responsibility, however, continues to reside with the person(s) with the authority to effect change.

The requirement to assure that deviations reported by the quality assurance unit (QAU) are communicated to the study director and that corrective actions are taken and documented, does not mean that management itself must communicate the findings and take appropriate corrective action. An efficient QAU will document deviations and the fact that corrective action has already occurred in reports which are distributed both to management and to the study director. The need for additional management follow-up will then be necessary only in those few instances where corrective action was not adequately negotiated between the QAU and the scientific staff before the issuance of the QAU report. When corrective action is underway but not complete at the time of the QAU report, the report need only indicate that fact with additional follow-up provided in subsequent reports.

§ 58.33 Study Director

For each nonclinical laboratory study, a scientist or other professional of appropriate education, training, and experience, or combination thereof, shall be identified as the study director. The study director has overall responsibility for the technical conduct of the study, as well as for the interpretation, analysis, documentation and reporting of results, and represents the single point of study control. The study director shall assure that:

(a)   The protocol, including any change, is approved as provided by § 58.120 and is followed.

The study director does not approve the protocol but only makes certain that approval is obtained from sponsor management.

    (b)   All experimental data, including observations of unanticipated responses of the test system are accurately recorded and verified.

The study director is not required to observe every data collection event, but should assure that data is collected as specified by the protocol and the standard operating procedures and that data collection includes the accurate recording of unanticipated responses of the test system. The study director should also review data periodically, or assure that such review occurs, to promote the accurate recording of data and to assure that data are technically correct.

    (c)   Unforeseen circumstances that may affect the quality and integrity of the nonclinical laboratory study are noted when they occur, and corrective action is taken and documented.

Systems must be in place to assure that the study director is promptly notified of unforeseen circumstances that may have an effect on the integrity of the study. The study director must then assure that corrective action is taken and documented in response to those unforeseen circumstances.

    (d)   Test systems are as specified in the protocol.

The determination of the appropriateness of the test system is a scientific decision made by management at the time of protocol approval. The study director need only assure that protocol specifications are followed.

    (e)   All applicable good laboratory practice regulations are followed.

This section suggests the need for frequent interaction between the study director and quality assurance personnel. Deviations from GLP requirements noted by the QAU must be reported periodically to management and to the study director. If those reports indicate that corrective action is still needed for any deviation from regulatory requirements, it is the study director's responsibility to assure that corrective action occurs.

The study director's role is not simply to react to reports of regulatory deviations from the QAU, but also to play a proactive role to assure that study personnel are aware of GLP requirements and that deviations from those requirements do not occur.

(f) All raw data, documentation, protocols, specimens, and final reports are transferred to the archives during or at the close of the study.

Materials may be transferred to the archives as the study progresses or at the close of the study. Although FDA has defined "the close of the study" as the time when the final report of the study is signed by the study director, it will not be a violation of regulatory requirements if materials reach the archives a reasonable period of time after the signature date.

§ 58.35 Quality Assurance Unit

(a) A testing facility shall have a quality assurance unit which shall be responsible for monitoring each study to assure management that the facilities, equipment, personnel, methods, practices, records, and controls are in conformance with the regulations in this part. For any given study the quality assurance unit shall be entirely separate from and independent of the personnel engaged in the direction and conduct of that study.

Arguments for maintaining a full-time staff of quality assurance professionals have been previously delineated in the discussion of the definition of "quality assurance unit."

(b)   The quality assurance unit shall:

(1)   Maintain a copy of a master schedule sheet of all nonclinical laboratory studies conducted at the testing facility indexed by test article and containing the test system, nature of study, date study was initiated, current status of each study, identity of the sponsor, and name of the study director.

FDA believes that maintenance of a detailed master schedule sheet is essential to the proper functioning of the quality assurance unit. In actual practice few quality assurance groups use the master schedule in the performance of quality assurance functions. Few do more than maintain a master schedule for the benefit of FDA inspectors who use it to gauge the volume of GLP-regulated work being conducted by a laboratory and to aid in the random selection of studies for review during an inspection.

There is no requirement for the QAU to prepare the master schedule. The master schedule may be prepared by some organizational unit other than the QAU as long as the QAU maintains a copy in its files.

FDA has indicated that a study should first appear on the master schedule on the date the protocol is signed by the study director. A study may come off the master schedule when the final report of the study is signed by the study director.

In the preamble (¶¶ 8 and 13) to the 1987 GLP revisions (4) the master schedule was referred to as "raw data." In a subsequent clarification, Dr. Paul Lepore indicated that the term "raw data" had appeared in quotes in the preamble to indicate that the term was not being used as defined in § 58.3(k) of the regulations. Rather the term was used to emphasize that copies of the master schedule were subject to the record retention requirements of §§ 58.190 and 58.195.

Additional language in the preamble (¶ 15) to the GLP revisions of 1987 (4) as well as enforcement policies of individual FDA investigators have broadly interpreted the requirement to include the "current status of each study" on the master schedule. According to this view, the master schedule should include such study events as test article-mixture preparation, test system dosing, in life observation, etc. Because such detailed information is usually available in other study documentation (e.g., protocol, study schedules, etc.), most laboratories limit a description of "current status" to broad categories such as "in life phase," "study terminated," "report preparation," and "report issuance."

It is permissible to identify the sponsor on the master schedule by code rather than name. This allows a contract laboratory to protect client confidentiality if the master schedule is examined by one of many clients. The contract laboratory must, however, make the names of sponsors available to FDA upon request.

Many laboratories maintain the master schedule on computer, and find it a helpful tool for the allocation of resources and the scheduling of work. A computerized master schedule can also provide the index of archive materials required by § 58.190(e) of the regulations.

> (2) Maintain copies of all protocols pertaining to all nonclinical laboratory studies for which the unit is responsible.

A proper discharge of QAU responsibilities requires a knowledge of protocol requirements. One of the QAU's responsibilities is to inspect study conduct to assure that there are no deviations from protocol requirements. Preparation for and conduct of those inspections requires ready access to a copy of the protocol and all protocol amendments.

> (3) Inspect each nonclinical laboratory study at intervals adequate to assure the integrity of the study and maintain written and properly signed records of each periodic inspection showing the date of the inspection, the study inspected, the phase or segment of the study inspected, the person performing the inspection, findings and problems, action recommended and taken to resolve existing problems, and any scheduled date for reinspection. Any problems found during the course of an inspection which are likely to affect study integrity shall be brought to the attention of the study director and management immediately.

"Inspect" has been defined by FDA to mean an actual examination and direct observation of the facilities and operations for a given study while the study is in progress and not merely a review of the records of a study. The QAU function is to observe and report on the state of compliance of a study with the requirements of the study protocol, laboratory standard operating procedures, and the GLP regulations. The QAU role is not just to verify the results of a study.

Each QAU may exercise reasonable flexibility and judgment to establish an inspection schedule which it believes is "adequate to assure the integrity of the study." However, FDA has indicated that every study must be inspected in-process at least once. Additional inspections may be randomly scheduled such that over a series of studies each phase for each type of study is inspected. Any random sampling approach to inspections should be statistically based and should be described and justified in the QAU's standard operating procedures.

Under U.S. Department of Agriculture animal welfare regulations [15], nonclinical laboratory studies in animals must be reviewed and approved by the testing facility's Institutional Animal Care and Use Committee (IACUC). FDA has indicated that IACUC review is part of the conduct of a nonclinical laboratory study and, therefore, IACUC activities should be periodically inspected by QAU. Because IACUC activities are subject to QAU inspection, FDA has indicated that a member of the QAU may not serve as a voting member of the IACUC but may serve as a non-voting member.

The information to be recorded in QAU inspection records is straightforward, as is the requirement for the QAU to immediately report significant problems to management and the study director.

> (4) Periodically submit to management and the study director written status reports on each study, noting any problems and the corrective actions taken.

The frequency of the QAU's periodic reports to management is left to the discretion of the laboratory. Reports at approximate monthly intervals are fairly standard within the regulated community. The description of problems noted during QAU inspections need not be extremely detailed unless the problems remain uncorrected. The primary purpose of the report is to assure management that study quality is being maintained and that management intervention is not required.

(5) Determine that no deviations from approved protocols or standard operating procedures were made without proper authorization and documentation.

As noted previously, QAU review of adherence to protocol requirements and standard operating procedures (SOP) are part and parcel of the inspection process.

(6) Review the final study report to assure that such report accurately describes the methods and standard operating procedures, and that the reported results accurately reflect the raw data of the nonclinical laboratory study.

The QAU audit should verify the accuracy and completeness of data and information presented in the final report. The audit should include the narrative description of materials, methods, and results as well as all tabulated data.

For critical study data (e.g., microscopic pathology data and data on tumor incidence) a QAU may elect to perform a 100% audit. For other data a random sampling approach to the audit is perfectly acceptable. Any such random sampling program should be statistically based (16, 17).

For reasons described in the discussion of § 58.185(c), the QAU will normally audit the final draft of the report before it is signed by the study director.

(7) Prepare and sign a statement to be included with the final study report which shall specify the dates inspections were made and findings reported to management and to the study director.

The list of inspection dates in the QAU statement may not be sufficient to reveal the extent of QAU audit and inspectional activity for any given study (e.g., when several inspections of a study occur on the same date). For this reason, some laboratories also list the study phases which were inspected even though this is not required by the regulations.

When a random sampling approach to the inspection process is used, it may be desirable to indicate the date(s) of inspection(s) of similar studies during a period which includes the time of conduct of the study for which the QAU statement is being prepared. Any such additional inspections should be clearly labeled as such.

    (c)   The responsibilities and procedures applicable to the quality assurance unit, the records maintained by the quality assurance unit, and the method of indexing such records shall be in writing and shall be maintained. These items including inspection dates, the study inspected, the phase or segment of the study inspected, and the name of the individual performing the inspection shall be made available for inspection to authorized employees of the Food and Drug Administration.

The Quality Assurance SOP manual should describe QAU audit and inspection techniques with attached inspection checklists, if used. Statistically based methods for random selection of phases of studies for inspection and for random selection of data points during final report audits should be described and justified. Any designation of study phases as "critical" or "noncritical" used to establish the frequency of study inspections should also be described and justified. The SOP manual should also describe the method for communicating audit and inspection findings to the study director and management, including a definition of who receives a copy of the reports. Lastly, the SOP manual should describe QAU record filing systems and the method for indexing those records. For filing and indexing systems, the QAU will find it most efficient to base its filing system on the study numbering system used by the safety testing laboratory. It can then utilize the safety laboratory's archive index system for indexing QAU records. The indexing system for QAU records should permit speedy access to such records in the event of any FDA request to review those records during an FDA inspection. FDA may review and copy any QAU records except those excluded by § 58.15(a).

(d)   A designated representative of the Food and Drug Administration shall have access to the written procedures established for the inspection and may request testing facility management to certify that inspections are being implemented, performed, documented, and followed-up in accordance with this paragraph.

(Collection of information requirements approved by the Office of Management and Budget under number 0910-0203)

As previously mentioned, QAU records of findings and problems and of corrective actions recommended and taken are exempt from routine FDA inspection. To compensate for this lack of routine inspectional authority, the FDA has access to the QAU's written procedures. FDA may review QAU written procedures to judge the adequacy of inspection schedules and to determine whether systems are in place for communicating inspection findings to management personnel. FDA may also request facility management to certify, in writing, that inspections are being implemented, performed, documented, and followed up in accordance with GLP requirements.

SUBPART C -- FACILITIES

§ 58.41   General

Each testing facility shall be of suitable size and construction to facilitate the proper conduct of nonclinical laboratory studies. It shall be designed so that there is a degree of separation that will prevent any function or activity from having an adverse effect on the study.

If a testing facility is too small to handle the volume of work which it has set out to do, there may be an inclination to mix incompatible functions. Examples might include the simultaneous conduct of studies with incompatible species (e.g., old world primates and new world primates) in the same room, setting up a small office in the corner of an animal housing area, housing an excessive number of animals in a room, or storing article/carrier mixtures in an animal room.

The facility should be constructed of materials which facilitate cleaning. Heating, ventilation, and air conditioning (HVAC) systems should be of adequate capacity to produce environmental conditions which comply with employee and animal health and safety standards and should be designed to prevent cross-contamination.

The location of a facility (e.g., next to a farm where pesticides, herbicides, and fertilizers are frequently used or next door to a chemical factory which generates noxious fumes) could have an adverse effect on the conduct of a nonclinical laboratory study unless the facility is designed to protect against outside environmental contaminants. Although the GLP revisions of 1987 eliminated "location" as a consideration in § 58.41, location is still a strong consideration in the design and construction of nonclinical laboratories.

Facilities should be designed to avoid disturbances such as intermittent or continuous noise from within or outside the facility, frequent traffic in and out of animal rooms, obnoxious odors (e.g., chemical odors which are carried by ventilation systems from laboratories to animal housing areas), inadequate or poorly diffused illumination, and similar disturbances which can be caused by facility design factors. Extreme care should be taken to design special protection for those types of animals (e.g., pregnant animals) that are especially sensitive to interfering disturbances.

In short, FDA is concerned that a facility be designed and constructed to ensure the adequacy of the facility for conducting nonclinical laboratory studies and to ensure the quality and integrity of study data.

§ 58.43 Animal Care Facilities

(a) A testing facility shall have a sufficient number of animal rooms or areas, as needed, to assure proper: (1) Separation of species or test systems, (2) isolation of individual projects, (3) quarantine of animals, and (4) routine or specialized housing of animals.

Note the words "as needed" and "proper." The facility's veterinarian-in-charge should be consulted as to when generally accepted standards for laboratory animal care require the separation, isolation, or specialized housing of animals. It is generally accepted that all newly received animals should undergo a quarantine and acclimation period.

"Isolation" generally connotes a setting apart, by use of physical barriers, from all other projects. "Separation," on the other hand, can be accomplished by spatial arrangements (e.g., two projects can be assigned to different parts of the same room).

(b)   A testing facility shall have a number of animal rooms or areas separate from those described in paragraph (a) of this section to ensure isolation of studies being done with test systems or test and control articles known to be biohazardous, including volatile substances, aerosols, radioactive materials, and infectious agents.

A laboratory involved in work with the hazardous materials described in this section also needs to be familiar with regulations of the Occupational Safety and Health Administration (or state equivalent), the Department of Agriculture and the Nuclear Regulatory Commission, all of which have a role in the regulation of such materials.

(c)   Separate areas shall be provided, as appropriate, for the diagnosis, treatment, and control of laboratory animal diseases. These areas shall provide effective isolation for the housing of animals either known or suspected of being diseased, or of being carriers of disease, from other animals.

If a laboratory's policy is to euthanize all diseased animals, it need not provide separate areas for the diagnosis, treatment, and control of laboratory animal diseases. Even if the laboratory does not have such a policy, there may be instances (e.g., non-contagious diseases) where diseased animals need not be isolated for treatment. Whether or not to treat and whether or not to isolate is a scientific decision which should be made by the study director in consultation with other scientific personnel.

If a laboratory intends to treat rather than euthanize diseased animals, it is best to have an area, separate from other animal housing and holding areas, for the isolation of diseased animals (if this is deemed necessary). A second, separate area may be needed to treat animals with contagious diseases separate from those animals being treated for noncontagious diseases.

> (d) When animals are housed, facilities shall exist for the collection and disposal of all animal waste and refuse or for safe sanitary storage of waste before removal from the testing facility. Disposal facilities shall be so provided and operated as to minimize vermin infestation, odors, disease hazards, and environmental contamination.

A laboratory may dispose of animal waste and refuse on-site (e.g., incineration) or may use a contract service for pick-up and disposal. Some animal waste and refuse may meet EPA's definition of "hazardous waste" (e.g., waste or refuse from animals treated with hazardous materials or animals carrying infectious diseases) and must be disposed of in compliance with EPA regulations issued under the Resource Conservation and Recovery Act. Waste and refuse from animals treated with radioactive materials must be disposed of in compliance with regulations of the Nuclear Regulatory Commission.

Containers with tight-fitting lids should be used for the temporary storage of animal waste and refuse before disposal to minimize vermin infestation, odors, disease hazards, and environmental contamination.

§ 58.45 Animal Supply Facilities

> There shall be storage areas, as needed, for feed, bedding, supplies, and equipment. Storage areas for feed and bedding shall be separated from areas housing the test systems and shall be protected against infestation or contamination. Perishable supplies shall be preserved by appropriate means.

Animal feed and bedding should never be stored in areas where animals are housed. It is also contrary to good animal husbandry practices to store supplies and equipment in animal housing areas.

Animal feed and bedding should be stored off the floor to facilitate cleaning. Food storage areas and areas used to store other perishable supplies should be temperature-controlled to protect against deterioration of the stored materials.

The first line of defense against vermin should be perimeter control, that is, controls to prevent the entry of vermin into a facility. If vermin control within the facility is necessary, care should be taken to protect supplies of feed and bedding from contamination by vermin control materials.

§ 58.47 Facilities for Handling Test and Control Articles

(a)   As necessary to prevent contamination or mix-ups, there shall be separate areas for:

(1)   Receipt and storage of the test and control articles.

(2)   Mixing of the test and control articles with a carrier, e.g., feed.

(3)   Storage of the test and control article mixtures.

(b)   Storage areas for the test and/or control article and test and control mixtures shall be separate from areas housing the test systems and shall be adequate to preserve the identity, strength, purity, and stability of the articles and mixtures.

The twin goals of § 58.47 are to prevent cross-contamination and mix-ups. Facility management must provide the necessary degree of separation to meet these goals. Separate rooms are not required for each of the described functions if adequate separation can be provided by spatial arrangements within a room, by special air-handling techniques, and/or by strictly enforced procedural requirements.

Dedicated areas are usually provided for the receipt and storage of test and control articles. Such articles are usually stored under lock and key. Areas for weighing test and control articles are often equipped with special air-handling systems, sometimes room-wide and other times limited to the area immediately surrounding the weighing devices. Many laboratories have a policy for weighing only one test or control article at any one time in any one area.

Operations with high cross-contamination potential (e.g., mixtures of test or control articles with animal diets) are often conducted in small, dedicated, individual cubicles equipped with special and separate air handling systems or are conducted under a fume hood. Special mixing equipment (e.g., enclosed twin-shell blenders) can be used to reduce the chance of cross-contamination.

If it is necessary to store test and control article mixtures, such materials should be stored entirely separate from animal housing areas. Special storage conditions (e.g., refrigeration and protection from light) must be available if needed to preserve and maintain the quality and stability of the mixtures.

§ 58.49 Laboratory Operation Areas

> Separate laboratory space shall be provided, as needed, for the performance of the routine and specialized procedures required by nonclinical laboratory studies.

A laboratory must provide adequate and, if necessary, separate space for the performance of routine and specialized procedures. Examples of specialized procedures include aseptic surgery, necropsy, histology, radiography, handling of biohazardous materials, and cleaning and sterilizing of equipment and supplies.

§ 58.51 Specimen and Data Storage Facilities

> Space shall be provided for archives, limited to access by authorized personnel only, for the storage and retrieval of all raw data and specimens from completed studies.

A laboratory which conducts nonclinical laboratory studies must provide space for the storage of raw data and specimens from such studies. Access to the archives must be controlled. This is best accomplished by providing a lockable area and by defining in the laboratory's standard operating procedures who has access to archive materials and under what conditions (e.g., use only within the archives or "check-out" rights).

Raw data and specimens need not be transferred to the archives until completion of the study. However, many laboratories elect to transfer material to the archives as it is completed to provide greater data security. FDA has stated that all materials must be transferred to the archives within a reasonable period of time after the study director signs the final report.

See the discussion of § 58.190 for other archive requirements.

SUBPART D -- EQUIPMENT

§ 58.61 Equipment Design

> Equipment used in the generation, measurement, or assessment of data and equipment used for facility environmental control shall be of appropriate design and adequate capacity to function according to the protocol and shall be suitably located for operation, inspection, cleaning, and maintenance.

Equipment used to generate, measure, or assess data should undergo a validation process to ensure that such equipment is of appropriate design and adequate capacity and will consistently function as intended. Examples of such equipment include scales; balances; analytical equipment (HPLC, GC, etc.); hematology, blood chemistry, and urine analyzers; computerized equipment for the direct capture of data; and computers for the statistical analysis of data. Because the data generated, measured, or assessed by such equipment are the essence of a nonclinical laboratory study, the proper functioning of such equipment is essential to valid study results.

Safety assessment scientists and technicians and even quality assurance personnel sometimes overlook the importance of environmental control equipment to valid study results. Animals stressed by extremes of temperature or humidity may yield spurious data; reproductive toxicology studies may be compromised by malfunctioning timers for the control of light/dark cycles; inadequate air filtration may expose experimental animals to environmental contaminants which confound experimental results.

All equipment described in § 58.61 should be located in such manner as to promote proper operation, inspection, cleaning, and maintenance.

§ 58.63 Maintenance and Calibration of Equipment

    (a) Equipment shall be adequately inspected, cleaned, and maintained. Equipment used for the generation, measurement, or assessment of data shall be adequately tested, calibrated and/or standardized.

The need for regular inspection, cleaning, and maintenance of equipment is well recognized in the scientific community. A laboratory should establish schedules for such operations based on manufacturer's recommendations and laboratory experience. In most instances these schedules will be defined as to periodicity, although in some cases an "as needed" schedule will be acceptable.

The terms "test," "calibration," and "standardization" are interrelated. Each term has a special meaning, but there is some overlap of the terms.

"Test" can be defined as an examination of an item or system to determine compliance with its specifications. Under this definition "test" would include operations to "calibrate" or "standardize" but would also include the process of total system validation.

"Calibration" has been defined as a comparison of a measurement standard or instrument of known accuracy with another standard or instrument to detect, correlate, report, or eliminate by adjustment, any variation in the accuracy of the item being compared (18).

"Standardization" is a comparison with a standard of known and accepted value. Standards may be of several sources: primary standards (prototype state of the art standards found at NIST, the National Institute of Standards and Technology (formerly known as the National Bureau of Standards (NBS)), or national equivalent outside of the U.S.); secondary, working standards (standards calibrated to primary standards, which are used for working tools and instruments); and in-house developed or interim standards (standards developed and used by a particular facility when no primary standard is available).

Scales and balances should be calibrated at regular intervals, usually ranging from 1 to 12 months, depending on manufacturers' recommendations, laboratory experience, and the extent of use. Intervals should be selected with a recognition that if a scale or balance is found to be out of calibration, it will cast doubt on the accuracy of every weight measured by that scale or balance since the last calibration. Scales and balances should also be standardized with a range of standard weights at frequent intervals. Many laboratories standardize scales and balances before each use, and some also standardize at periodic intervals during each use. The range of standard weights should bracket the expected experimental values. Standard weights should be traceable to NIST standards and should, themselves, be periodically calibrated.

The use of standard solutions, reference standards, and quality control samples, whether prepared by the laboratory or purchased commercially, is essential to valid analyses of test and control article/carrier mixtures and biological fluids (blood, serum, plasma, etc).

A Ph meter should be standardized before each use according to directions in the manufacturer's manual.

Electrocardiographs usually have a built-in facility for generating an electrical impulse of known intensity. This facility should be used during the recording of electrocardiograms to check periodically on the proper functioning of the equipment.

Heating, ventilation, and air conditioning (HVAC) equipment should be regularly inspected and maintained. Filters on environmental control equipment should be inspected on a regular basis and changed as needed.

(b) The written standard operating procedures required under § 58.81(b)(11) shall set forth in sufficient detail the methods, materials, and schedules to be used in the routine inspection, cleaning, maintenance, testing, calibration and/or standardization of equipment, and shall specify, when appropriate, remedial action to be taken in the event of failure or malfunction of equipment. The written standard operating procedures shall designate the person responsible for the performance of each operation.

All aspects of a laboratory's program for the routine inspection, cleaning, maintenance, testing, calibration and/or standardization of equipment must be in writing (i.e., SOPs supplemented, as necessary, by equipment manuals). This would include a description of cleaning materials; inspection, cleaning, and maintenance methods and schedules; calibration and standardization methods and parameters; and the job title of personnel responsible for the performance of each operation. Specification of remedial actions to be taken in response to equipment failure or malfunction should be as comprehensive as possible. Common trouble-shooting problems with appropriate remedial action are frequently included in equipment manufacturers' manuals which can be cited in the standard operating procedures. For other types of problems, it will generally be sufficient to indicate in the standard operating procedures that professional assistance will be enlisted (e.g., manufacturers' repair services).

Copies of equipment standard operating procedures must be easily and readily accessible by laboratory personnel.

The phrase "when appropriate" means that a laboratory only need specify remedial action in response to equipment failure or malfunction when remedial action is appropriate to the piece of equipment. A laboratory may elect to discard rather than repair faulty equipment; however, records for the discarded equipment, including records of previous maintenance, calibration, etc. must be retained for the length of time described in § 58.195(b) and (f).

(c)   Written records shall be maintained of all inspection, maintenance, testing, calibrating and/or standardizing operations. These records, containing the date of the operation, shall describe whether the maintenance operations were routine and followed the written standard operating procedures. Written records shall be kept of non-routine repairs performed on equipment as a result of failure and malfunction. Such records shall document the nature of the defect, how and when the defect was discovered, and any remedial action taken in response to the defect.
(Collection of information requirements approved by the Office of Management and Budget under number 0910-0203)

As with any activity required by regulation, records must be maintained of all equipment inspection, maintenance, testing, calibrating and/or standardizing operations. The records required by this section of the regulations are necessary to the reconstruction of a study and provide the FDA with added assurance as to the validity and integrity of data. FDA has indicated, however, that it is not necessary to maintain records of cleaning operations on the theory that the costs of maintaining such records exceeds the benefits.

Records of routine maintenance operations may reference the standard operating procedures for a description of the operations. For non-routine repairs in response to equipment failure or malfunction, repair records must contain the following detailed information: nature of the defect, how the defect was discovered, when the defect was discovered, and remedial action taken in response to the defect. Remedial action should include a review of possible effects on data generated before the defect was discovered. Because repairs are likely to involve repairmen from outside the laboratory, care must be taken to ensure that such persons provide full documentation of the nature of the problem and remedial action taken in response to the problem.

Equipment inspection, maintenance, and repair records can be recorded in a log book especially designed for that purpose. For equipment that is moved from laboratory to laboratory, the log book should accompany the equipment when it is moved. Documentation of calibrating or standardizing operations, on the other hand, may be more efficiently recorded with the associated records of the data acquisition activities.

SUBPART E -- TESTING FACILITIES OPERATION

§ 58.81 Standard Operating Procedures

(a)   A testing facility shall have standard operating procedures in writing setting forth nonclinical laboratory study methods that management is satisfied are adequate to insure the quality and integrity of the data generated in the course of a study. All deviations in a study from standard operating procedures shall be authorized by the study director and shall be documented in the raw data. Significant changes in established standard operating procedures shall be properly authorized in writing by management.

Preparation of written standard operating procedures (SOPs) was a major undertaking for most GLP-regulated laboratories. Keeping SOP manuals up to date continues to be a major effort for these labs. To assure that SOP manuals remain up to date, many laboratories have a policy for mandatory, periodic review (and update, if necessary) of all standard operating procedures.

Study protocols define "what" is to be done during the course of a study. SOPs define "how" to carry out protocol-specified activities.

There are many acceptable formats for SOPs. The author prefers an activity-oriented format, written in playscript style and including a designation of the actor (i.e., who is responsible for the activity) and a chronological listing of action steps (i.e., of what the activity consists) (19). See Figure 1 for a sample of an SOP in this format. Prime consideration should be given to making the SOP manual user-friendly so that it is a document which invites rather than discourages routine usage by those responsible for performing tasks in compliance with the SOP.

In the preparation and revision of SOPs, a major consideration is the degree of detail to be incorporated into SOPs. As a general rule, SOPs should be detailed enough to provide meaningful direction to study personnel for the conduct of routine laboratory activities. In determining the level of detail, it is acceptable to take into consideration the education, training, and experience of the personnel who will be responsible for those activities. For example, an analytical procedure to be carried out by a trained chemist would instruct the chemist to pipette 5 ml of a reagent, but need not provide detail of how to pipette. It is generally not advisable to specify suppliers of materials in SOPs because suppliers may change frequently. It is always advisable to allow for a range of acceptable approaches to any procedure if a more specific, restrictive, and defined activity is not necessary to assure study quality. If written too restrictively, SOPs are frequently in need of revision. On the other hand, if insufficient detail is included in the SOPs, they fail to provide

Figure 1 Sample SOP

PARKE-DAVIS PHARMACEUTICAL RESEARCH DIVISION	PROCEDURES	VOLUME: SECTION: 80.745.05 80.745
DEPARTMENT OF PATHOLOGY AND EXPERIMENTAL TOXICOLOGY		EXHIBIT:
SUBJECT: TEST AND CONTROL ARTICLES - STABILITY RE-ANALYSIS		

PURPOSE

The following provisions elaborate upon the test and control article policy (80.745.02), establishing specific responsibilities and delegations of authority involved in the periodic re-analysis of test and control articles for assurance of stability.

LOCATION AFFECTED

Pathology and Experimental Toxicology, Ann Arbor, Michigan.

PROCEDURE

<u>RESPONSIBILITY</u>   <u>ACTION</u>

Study Director or
Scientist/Technician
Designee

1. Review the available stability data of test and control articles prior to initiation of a study. Arrange for periodic re-analysis of each lot if the stability of the test or control article cannot be determined before initiation of a study or available stability data does not cover the projected time period of the study.

2. At termination of studies of less than 6 months' duration (including genetic toxicology and reproductive toxicology investigations), select a sample (approximately 200 mg if possible) of test or control article for potency re-analysis.

<u>NOTE:</u>   Samples for re-analysis should be taken from the container in use at the time.

3. If the study is to last more than 6 months, forward sample for re-analysis at least every 6 months.

<u>NOTE:</u>   If studies with the same test and control articles are conducted concurrently, <u>ONE POTENCY RE-ANALYSIS SAMPLE COLLECTED AFTER COMPLETION OF THE LAST STUDY WILL SUFFICE.</u> Please refer to Drug Inventory Forms for timing of sample collection.

4. If at any time, regardless of study duration, a lot is nearing depletion, or a lot is being returned and a potency assay has not been conducted, submit a sample for re-analysis.

5. Receive the written report, evaluate stability test results and incorporate document into the appropriate study file(s).

6. If a significant loss of activity or increase in contaminant(s) has occurred, submit a sample for re-test and notify Departmental Supervision and Quality Assurance immediately (significance of potency loss will vary according to the precision of the assay procedure, but a loss of $\geq 10\%$ is cause for concern; in general, no single contaminant should be present at $>0.5\%$ and total contaminants should be $\leq 2\%$).

Directors or Section Heads,
Pathology and Experimental
Toxicology and Quality
Assurance Representative

7. If re-test confirms the loss of potency or increase in contaminants, determine how this information affects the associated studies.

DATE: [Effective Date Entered Here]	[Management Approval Signature]	PAGE 1 of 1

adequate direction to study personnel.   With experimentation and experience a laboratory can strike a reasonable balance between too much and not enough detail. It is always a good idea to solicit comments from those who use the SOP manual (the workers at the bench) in striking this balance.

If an exception to a standard operating procedure is to be made for an individual study, that exception must be authorized in writing by the study director, and the written authorization must be maintained with the raw data for the study. If a change in procedure represents a new, standard way of doing things, then the SOP should be revised, and the revision approved (e.g., by signature) by laboratory management.

    (b)   Standard operating procedures shall be established for, but not limited to, the following:

    (1)   Animal room preparation.

    (2)   Animal care.

    (3)   Receipt, identification, storage, handling, mixing, and method of sampling of the test and control articles.

    (4)   Test system observations.

    (5)   Laboratory tests.

    (6)   Handling of animals found moribund or dead during study.

    (7)   Necropsy of animals or postmortem examination of animals.

    (8)   Collection and identification of specimens.

    (9)   Histopathology.

   (10)   Data handling, storage, and retrieval.

   (11)   Maintenance and calibration of equipment.

   (12)   Transfer, proper placement, and identification of animals.

As suggested by the phrase "but not limited to," the list of SOP topics in § 58.81(b) should be considered illustrative, not comprehensive. Many of the topics (e.g., laboratory tests) might involve a hundred or more individual SOP titles. The range of topics for which SOPs are required will be governed by the variety of studies routinely conducted in the laboratory. For each procedure required by each type of study, the laboratory should prepare an SOP describing how that procedure should be performed. If a study activity is not yet "standard" or is intended to be a one-time event, it is acceptable to incorporate a detailed description of the "how-to" for that activity in the study protocol or in a laboratory notebook. If such activities become routine, however, an SOP should be prepared.

In the foregoing discussion of § 58.35(b)(3) it was indicated that FDA considers Institutional Animal Care and Use Committee (IACUC) review of nonclinical laboratory studies in animals to be part of the conduct of those studies. FDA has also indicated that IACUC functions should be described in SOPs (even though the U.S. Department of Agriculture regulations (15) which mandate IACUC review do not require that IACUC functions be described in written SOPs!). FDA has specifically stated (20) that IACUC SOPs should include the following:

- A document from a high-ranking laboratory official which states that the laboratory does not condone or support inhumane treatment of animals and that it is the policy of the laboratory to maintain, hold, and use animals in compliance with all applicable regulations, guidelines, and policies;

- A description of committee members (including the chair), the number of members and their terms of office, and the procedure for replacing committee members;

- A definition of a quorum for committee activities;

- A description of how the committee makes decisions;

- A description of committee documents, including what items go out with a meeting agenda and what items should be described in meeting minutes.

Some laboratories establish a hierarchy of documents and specify that SOPs describe the approved method for study conduct unless an alternate methodology is described in a study protocol. In such a case, the alternate methodology would only be applicable for a study where the protocol so provides. Because the study director must sign the protocol, such a system provides an easy method for compliance with § 58.81(a) which requires the study director to authorize all deviations in a study from standard operating procedures.

(c)   Each laboratory area shall have immediately available laboratory manuals and standard operating procedures relative to the laboratory procedures being performed. Published literature may be used as a supplement to standard operating procedures.

If SOPs are to provide guidance to study personnel on accepted methods for the conduct of routine study procedures, it follows that they must be readily available to the personnel performing those activities. SOPs should be available in or near the room where the activities will occur. The requirement for "immediately available" SOPs is not met if an employee must travel some distance in order to consult the SOP manual. In such a case the employee is more likely to guess, and perhaps guess wrongly, about the proper method for study conduct.

The entire SOP manual need not be "immediately available" as long as those SOPs which describe procedures to be performed are available.

Published literature (and manufacturers' equipment manuals) may supplement SOPs but will, as a general rule, not be an acceptable substitute for SOPs.

Well-prepared SOPs will serve as a good training tool for new employees and will provide a handy "crutch" for experienced personnel whose memory of study methods may need some refreshing. Although not a GLP requirement, many laboratories provide each employee with a complete copy of the SOP manual in addition to providing the mandatory "working copies" of individual SOP titles in each work area.

To meet the requirement for "standard" operating procedures, the laboratory is advised to develop a system to ensure that all working copies of the SOPs are identical. When SOP revisions are distributed, all holders of the manual should be instructed, at a minimum, to destroy the outdated version of the procedure. A better approach is to require the outdated version to be returned to and accounted for at a central location. Follow-up should then be provided to ensure the return of all copies of the outdated procedure. Ideally each distributed copy of the SOPs should be uniquely numbered, and employees should be instructed not to make copies of any individual SOP. This provides better control over the distribution process and helps ensure that all outdated versions of SOPs are destroyed.

With the advance in computer technology and the increased use of computer networks, many laboratories are making SOPs available in electronic form via read-only central computer files. Electronic SOPs help ensure that all personnel are using the current version of an SOP, reduce or eliminate the need for distribution of paper copies of the SOPs, and reduce or eliminate the need for follow-up to ensure that SOP manuals are updated properly. A master, hardcopy version of the SOPs which is authorized, signed, and dated by management still must be retained in the archives, and the historical file of SOPs should also contain hardcopy versions which have been authorized, signed, and dated by management. Electronic SOPs, just like hardcopy SOPs, must be readily available to study personnel.

> (d) A historical file of standard operating procedures, and all revisions thereof, including the dates of such revisions, shall be maintained.

The historical file of SOPs documents what standard operating procedures were in effect at any time during a laboratory's history. Because FDA inspection of a study often occurs years after the completion of that study, the historical file of SOPs will be of especial use to an FDA inspector. Including the effective date on the SOP itself will aid in maintenance of the historical file and will, also, make it easier to ascertain if any one SOP manual contains the current version of any individual SOP. Accessory documentation of effective dates (e.g., in the transmittal memo for the distribution of SOPs) is permissible but not recommended.

§ 58.83 Reagents and Solutions

> All reagents and solutions in the laboratory areas shall be labeled to indicate identity, titer or concentration, storage requirements, and expiration date. Deteriorated or outdated reagents and solutions shall not be used.

Good laboratory technique has always included proper labeling of reagents and solutions. Many laboratories provide supplies of standard labels which prompt laboratory personnel to include the four pieces of information mandated by the GLPs. "Identity" and "titer or concentration" present no problems. For "storage requirements" it is acceptable for laboratory SOPs to indicate that reagents and solutions may be stored at ambient room temperature unless otherwise indicated on the label. The standard label would then provide a space for "special storage conditions" (e.g., "refrigerate," "protect from light," etc.) The requirement to include an expiration date sometimes is resisted by laboratory personnel, especially for materials such as powder forms of histologic stains, crystalline sodium chloride, etc. For such materials there is no known expiration date, and it is acceptable to indicate "NONE" or "N/A" (not applicable) on the label for expiration date. The laboratory must, however, be prepared to justify this designation. For other materials an expiration date should always be indicated on the label. FDA has indicated that formal stability studies are not required to justify assigned expiration dates. It is sufficient to assign expiration dates based on literature references and/or laboratory experience.

The best guarantee that outdated reagents and solutions will not be used is a strictly enforced policy for discard of such materials, although that is not a GLP requirement. The GLPs require only that outdated materials not be used.

Official FDA enforcement policy requires adherence to GLP labeling requirements for all reagents and solutions in a laboratory where GLP-regulated work is conducted even if some of those reagents and solutions are used for work that is not GLP-regulated. FDA's concern is that reagents and solutions that are not adequately labeled, even if not intended for use in GLP-regulated studies, may have an adverse effect on laboratory work that is GLP-regulated.

§ 58.90 Animal Care

> (a)   There shall be standard operating procedures for the housing, feeding, handling, and care of animals.

This is simply a reiteration of the requirements of § 58.81.

(b) All newly received animals from outside sources shall be isolated and their health status shall be evaluated in accordance with acceptable veterinary medical practice.

Isolation is the separation of newly received animals from those already in the facility until the health of the newly received animals has been evaluated. Effective isolation minimizes the introduction of disease-causing agents into established animal colonies. It also allows time for the expression of clinical signs of disease which will permit culling of animals before they are placed on study.

Quality control by the animal vendor and a knowledge of the history of the animals are acceptable parts of an institution's isolation procedures. This information may limit the isolation period for rodents to the time necessary for inspection upon arrival; however, all newly received animals should be allowed a stabilization period prior to their use (21).

Isolation may occur in the same room where the study will be conducted. It is not necessary to provide separate, dedicated isolation areas. Laboratory personnel should solicit the expert advice of the veterinary staff in the establishment of isolation procedures.

(c) At the initiation of a nonclinical laboratory study, animals shall be free of any disease or condition that might interfere with the purpose or conduct of the study. If, during the course of the study, the animals contract such a disease or condition, the diseased animals shall be isolated, if necessary. These animals may be treated for disease or signs of disease provided that such treatment does not interfere with the study. The diagnosis, authorizations of treatment, description of treatment, and each date of treatment shall be documented and shall be retained.

Good science has always mandated the use of high quality, disease-free animals to reduce extraneous factors which might complicate the interpretation of experimental results.

The GLPs permit the treatment of diseases or conditions which develop during the course of a study. Any animal so treated should be isolated from other animals if necessary to protect against adverse effects on a study. Laboratories may elect to euthanize diseased animals rather than provide treatment.

If a laboratory elects to treat diseased animals, the GLPs specify documentation requirements for such treatment. These documentation requirements are straightforward and consistent with accepted veterinary medical practice.

> (d) Warm-blooded animals, excluding suckling rodents, used in laboratory procedures that require manipulations and observations over an extended period of time or in studies that require the animals to be removed from and returned to their home cages for any reason (e.g., cage cleaning, treatment, etc.) shall receive appropriate identification. All information needed to specifically identify each animal within an animal-housing unit shall appear on the outside of that unit.

There is no perfect system for identification of animals. Tattoos and color codes frequently fade and may need to be redone after 3 to 6 months. Toe clips and ear punches are occasionally obliterated by self-mutilation or mutilation by cagemates. Ear tags and collars fall off and need to be replaced. Cage cards can be lost or destroyed. Whatever system or combination of systems of animal identification are selected by a laboratory, the shortcomings of the selected system(s) must be recognized, and procedures must be developed to address those shortcomings.

Identification other than a cage card is not required for short-term studies where an animal is never taken from and returned to its cage during the course of a study. The preamble (¶ 35) to the GLP revisions of 1987 (4) also indicates that, when animals are housed individually in cages, cage cards plus detailed animal handling SOPs designed to prevent animal mix-ups will constitute an adequate animal identification system.

Any system of animal identification should provide an appropriate means for distinguishing one animal from all other animals housed in the same room. Each animal's identification only needs to be unique in the room where it is housed; it need not be unique to all studies ever conducted with that species in the laboratory.

Identification of suckling rodents might lead to cannibalization by the mother. Therefore, FDA has exempted suckling rodents from the identification requirements of this section. There are other, unique situations where placing identifying features on an animal has the potential of jeopardizing the validity of the study. One such type of study is the guinea pig sensitization study where metal ear tags, plastic collars, or the dyes in tattoos and other color markings may themselves produce a sensitization response; ear punching may produce inflammation that could jeopardize test results; and toe clipping may lead to excessive bleeding. As previously mentioned, a laboratory may elect to identify such animals by cage card only. However, the animals must be singly housed, and animal handling SOPs should provide specialized procedures for preventing animal mix-ups.

> (e) Animals of different species shall be housed in separate rooms when necessary. Animals of the same species, but used in different studies, should not ordinarily be housed in the same room when inadvertent exposure to control or test articles or animal mixup could affect the outcome of either study. If such mixed housing is necessary, adequate differentiation by space and identification shall be made.

Physical separation of animals by species is generally recommended to prevent interspecies disease transmission and to reduce anxiety due to interspecies conflict. In some situations it might be appropriate to house different species of rodents in the same room, such as when they are to be used for tests of the same test article and have a similar health status or when special containment is provided within rooms (e.g., laminar flow cabinets or filtered or microisolation cages). It is not uncommon for animals from one supplier to harbor microbial agents not found in animals of the same species from another supplier. Therefore, intraspecies separation is advisable when animals obtained from multiple sources differ in microbiological status (21).

The best rule is only one species from a single supplier in any one room and only one study per room. If mixed housing is absolutely necessary, the laboratory must provide adequate differentiation by space and identification and must take steps to minimize the possibility for disease transmission or cross-contamination.

> (f)   Animal cages, racks and accessory equipment shall be cleaned and sanitized at appropriate intervals.

The NIH *Guide* (21) recommends that animal cages be sanitized before use and, further, that solid-bottom rodent cages be washed once or twice a week and cage racks at least monthly. It is recommended that wire-bottom cages and cages for all other animals be washed at least every 2 weeks. Water bottles, sipper tubes, stoppers, other watering equipment, and feeders should be washed once or twice a week.

The rinse cycle for washing all equipment should use water of at least 82.2°C (180°F), or higher for a period long enough to ensure destruction of vegetative pathogenic organisms. Chemical treatment is an alternative method of disinfection. If chemicals are used, equipment should be rinsed free of chemicals prior to use. Periodic microbiologic monitoring is useful to determine the efficacy of disinfection or sterilization procedures.

> (g)   Feed and water used for the animals shall be analyzed periodically to ensure that contaminants known to be capable of interfering with the study and reasonably expected to be present in such feed or water are not present at levels above those specified in the protocol. Documentation of such analyses shall be maintained as raw data.

This GLP section was included as a result of FDA experience with toxicology studies of pentachlorophenol and diethylstilbestrol. In those studies the feeds used as carriers of the test article were found to contain varying quantities of pentachlorophenol and estrogenic activity. These contaminants invalidated the studies by producing erratic results.

Contaminant analysis of food and water for each and every study is not a requirement of § 58.90(g) nor is analysis for a laundry list of contaminants. What § 58.90(g) does require for every study is careful scientific consideration to determine whether there are any potential contaminants in the feed and water that are capable of interfering with test results. The study director and associated scientists from toxicology and other disciplines should consider each study in the light of its length, the expected toxicologic endpoints and pharmacologic activity of the test article, the test system, the route of administration, and other relevant factors to determine what contaminants could reasonably be expected to interfere. These considerations coupled with scientific literature, experience and anticipated levels of contamination should be used to determine which, if any, contaminants should be controlled and analyzed. FDA has said that it is unlikely that a blanket analysis conducted either by feed manufacturers or water authorities would be sufficient because such analyses would either provide data on contaminants which would not be expected to interfere or neglect to provide data for certain interfering contaminants.

Despite the foregoing, most labs rely on blanket analyses by feed manufacturers and water authorities, occasionally supplemented by analyses for a few additional contaminants also using a blanket approach (i.e., the same analyses for every study). It is likely that the type of scientific review expected by the FDA is simply not possible given the state of knowledge about test articles at the time safety studies are conducted.

Blanket analyses at least guard against the presence in the feed and water of known carcinogens (e.g., aflatoxins) which could interfere with the evaluation of a carcinogenicity study. Blanket analyses also assure that toxic materials (e.g., heavy metals, pesticides, coliform bacteria) will not compromise the results of longer term toxicity studies. Additional contaminant analyses should be conducted when the potential of interference by contaminants is known (e.g., tests for bivalent metal ions in the drinking water during the study of a tetracycline antibiotic and an analysis for estrogenic activity in the feed used during the study of an estrogen product).

The use of certified feeds for short-term studies is probably not justified unless a laboratory maintains only stocks of certified feeds to ensure that such feeds are used in longer term studies. Such a policy also eliminates the need to maintain inventories of two types of feed for each species of animal.

When analyzing the animals' drinking water for possible interfering contaminants, representative water samples should be drawn at the point of use by the animals to detect any possible contamination of the water by the delivery system.

Most laboratories describe their blanket analyses for contaminants in SOPs, which provide a full listing of the contaminants analyzed and acceptable levels for each. Study protocols in such cases merely make reference to the SOPs. If there is any analysis for contaminants not listed in the SOPs, the protocol should describe the additional contaminants and acceptable levels for each.

> (h)   Bedding used in animal cages or pens shall not interfere with the purpose or conduct of the study and shall be changed as often as necessary to keep the animals dry and clean.

Bedding should be absorbent, free of toxic chemicals or other substances that could injure animals or personnel, and of a type not readily eaten by animals. Bedding should be sufficient to keep animals dry between cage changes without coming into contact with watering tubes. Aromatic hydrocarbons from cedar and pine bedding materials can induce the biosynthesis of hepatic microsomal enzymes. Therefore, such beddings are not appropriate for use in nonclinical laboratory studies.

Bedding can be purchased that is guaranteed to be free of potentially interfering contaminants. In the absence of such a guarantee, the laboratory may wish to consider its own periodic analysis of bedding for contaminants.

Bedding used in cages or pens should be changed as often as is required to keep the animals dry and clean. For small rodents (e.g., rats, mice, and hamsters) one to three bedding changes per week will generally suffice. For larger animals (e.g., dogs, cats, and nonhuman primates) bedding should be changed daily.

Soiled bedding should be emptied from cages and pans under conditions that minimize exposure of animals and personnel to aerosolized waste.

> (i)   If any pest control materials are used, the use shall be documented. Cleaning and pest control materials that interfere with the study shall not be used.

The most effective pest control program prevents entry of vermin into a facility by screening openings, sealing cracks, and eliminating breeding and refuge sites. With the exception, perhaps, of boric acid or drying substances (e.g., silica gel), there are few pest control materials that are free of serious toxic properties. Therefore, the best policy is one which prohibits the use of toxic pesticides in rooms where animals are housed. If pest control materials are used in empty rooms, the room should not be used to house animals until the risk to animals has passed. This requires a knowledge of the degradation properties of the pesticide.

Application of pesticides must be recorded. The application must comply with federal, state, and local legal and regulatory requirements.

SUBPART F -- TEST AND CONTROL ARTICLE

§ 58.105 Test and Control Article Characterization

(a) The identity, strength, purity, and composition or other characteristics which will appropriately define the test or control article shall be determined for each batch and shall be documented. Methods of synthesis, fabrication, or derivation of the test and control articles shall be documented by the sponsor or the testing facility. In those cases where marketed products are used as control articles, such products will be characterized by their labeling.

The definition of "appropriate" characterization of test and control articles will vary depending on the stage of development of the articles. The amount of information on the first milligram quantity of material that is synthesized in the research laboratory will be much less than that available later in development when methods of synthesis have been scaled up to produce kilogram quantities. For test and control articles used in nonclinical laboratory studies, laboratory management should establish acceptable characteristics which are reasonably related to the stage of development.

Tests to characterize a test or control article as to its "identity" may be postponed until initial toxicology studies show a reasonable promise of the article's reaching the marketplace. FDA has indicated, however, that information on "strength" and "purity" should be available prior to the use of the article in a nonclinical laboratory study.

Methods of synthesis, fabrication, or derivation as well as identity (if established), strength, and purity characteristics of the material must be documented. Copies of this documentation must be included with study records and must be available for FDA inspection. In the case of contract testing facilities where the sponsor, for proprietary reasons, may not wish to release such information to the contract lab, the contract facility should have written assurance from the sponsor that such documentation exists.

Tests to establish the identity, strength, and purity of the test and control articles need not comply strictly with GLP requirements (e.g., protocol, QAU inspection requirements), but good documentation of analytical test results (usually in a laboratory notebook) and retention of raw data for such tests is a good practice. As the development process proceeds and the same material is used in both nonclinical and clinical studies, cGMP principles will apply to the production and characterization processes.

When marketed products are used as control articles, a copy of product labeling should be included with the study records.

(b)   The stability of each test or control article shall be deter-
      mined by the testing facility or by the sponsor either: (1)
      Before study initiation, or (2) concomitantly according to
      written standard operating procedures, which provide for
      periodic analysis of each batch.

In most cases the stability of test articles will not be established before the initiation of a study. In such cases laboratory SOPs should describe a policy for periodic reanalysis of each batch of the test article. Analytical methods for reanalysis must be stability indicating. The periodicity of the reanalysis is left to the discretion of the laboratory. Generally analyses are conducted at 3- to 6-month intervals during the period of test article use. In establishing the analysis interval the laboratory will want to weigh the risk of loss of a study because of test or control article instability against the costs of the periodic reanalyses.

The periodic stability reanalyses must be conducted in full compliance with the GLP regulations.

(c)   Each storage container for a test or control article shall be labeled by name, chemical abstract number or code number, batch number, expiration date, if any, and, where appropriate, storage conditions necessary to maintain the identity, strength, purity, and composition of the test or control article. Storage containers shall be assigned to a particular test article for the duration of the study.

Labeling requirements in § 58.105(c) are not controversial and are the minimum to ensure against mix-up of test or control articles. Expiration date needs to be included on the label only if one has been established. Some laboratories include a retest date on the label as a reminder of the need for periodic stability analyses. Only special storage conditions (e.g., "refrigerate," "protect from light," "protect from freezing," etc.) need to be included on the label.

In the preamble (¶ 38) to the 1987 GLP revisions (4), FDA declined to eliminate the storage container provision in § 58.105(c). Dr. Paul Lepore has indicated that ¶ 38 was referring only to the original storage container. According to the scenario envisioned by FDA, a lot of test or control article is selected for testing, characterized, and placed in a properly labeled storage container. This storage container must be retained for the duration of the test. Aliquots or samples of test article may be removed from this storage container and placed in intermediate "working" containers that are also properly labeled. However, these "working" containers need not be retained.

(d)   For studies of more than 4 weeks' duration, reserve samples from each batch of test and control articles shall be retained for the period of time provided by § 58.195.
(Collection of information requirements approved by the Office of Management and Budget under number 0910-0203)

FDA has indicated that "study initiation date" (defined in § 58.3(o)) and "study completion date" (defined in § 58.3(p)) are administrative dates and should not be used to determine whether or not a study is "of more than 4 weeks' duration." Instead, terms-of-the-art (e.g., 14-day acute study, 28-day repeated dose study, etc.) will determine whether reserve samples are required under § 58.105(d).

Reserve sample size should be at least twice the quantity necessary to perform all tests to determine whether the test or control article meets its established specifications for identity, strength, quality, purity, and stability. By retaining twice the quantity necessary to perform all tests, the laboratory will be able to supply a sample to the FDA, if requested, and still retain sufficient material to conduct its own tests.

§ 58.107 Test and Control Article Handling

Procedures shall be established for a system for the handling of the test and control articles to ensure that:

(a)   There is proper storage.

(b)   Distribution is made in a manner designed to preclude the possibility of contamination, deterioration, or damage.

(c)   Proper identification is maintained throughout the distribution process.

(d)   The receipt and distribution of each batch is documented. Such documentation shall include the date and quantity of each batch distributed or returned.

The general goals of § 58.107 are to maintain the integrity of and to provide accountability for the test and control articles throughout the period of use.

Integrity is maintained by ensuring that all containers of the articles are labeled properly, by storing all supplies of the articles in conformance with their labeling, and by ensuring that the articles are distributed, handled, and used in a manner that precludes the possibility of contamination, deterioration, or damage.

The accountability provisions of § 58.107(d) are met by records showing the date and quantity of test and control articles distributed from central stores for use in a study or series of studies and the date and amount of material returned to central stores at the end of a study or series of studies. To this should be added a system for documenting date and quantity for each use of a test or control article during the course of each study. A running inventory of test and control articles is not required but does provide an easy mechanism for periodically verifying the accuracy of test and control article usage.

§ 58.113 Mixtures of Articles with Carriers

    (a)   For each test or control article that is mixed with a carrier, tests by appropriate analytical methods shall be conducted:

    (1)   To determine the uniformity of the mixture and to determine, periodically, the concentration of the test or control article in the mixture.

    (2)   To determine the stability of the test and control articles in the mixture as required by the conditions of the study either (i) before study initiation, or (ii) concomitantly according to written standard operating procedures which provide for periodic analysis of the test and control articles in the mixture.

The requirements of § 58.113(a) substantially changed the "state of the art" for the conduct of nonclinical laboratory studies. Prior to the promulgation of GLP regulations, analytical tests to establish the homogeneity and stability of article/carrier mixtures were not routine nor were tests to determine the concentration of test and control articles in the mixtures used to deliver test and control articles to test systems.

There is no exemption from § 58.113(a) for short-term studies. If a study meets the definition of a "nonclinical laboratory study" all analytical requirements apply.

If a test or control article is administered in solution, homogeneity (uniformity) tests need not be conducted. For nonsolutions (e.g., suspensions and mixtures with diet), once the uniformity has been established for a given set of mixing conditions, it is not necessary to establish the uniformity of each subsequent batch that is mixed according to the same specifications. In taking samples for homogeneity testing, one must ensure that the samples are truly representative of the batch and that the total number of samples is adequate to prove uniformity. Typically samples are drawn from the top, middle, and bottom of the batch or according to a random sampling schedule. The number of samples from any one batch usually ranges from 6 to 9.

Stability of the article/carrier mixture can be established in conjunction with the homogeneity assays of nonsolutions. Separate stability tests will, of course, be required for solutions. Formal stability trials sufficient to show long-term stability of the mixtures are not required. Rather, stability should be established for a period which encompasses the period of use of the article/carrier mixture. "Period of use" should be defined as whichever of the following two time periods is longer: the time between preparation of the mixture and final administration of that mixture to the test system or the time between preparation of the mixture and the analysis of the mixture as required by § 58.113(a)(2). Often the period between preparation and analysis may be longer than the period between preparation and last administration to the test system.

Homogeneity and stability assays may be conducted before a study begins or may be conducted concurrent with the study. If the latter, poor assay results may, of course, result in invalidation of the study.

There are no established guidelines with regard to the frequency of periodic concentration assays. Some laboratories randomly select a sample from one concentration of article/carrier mixture per study per week. Other laboratories conduct an analysis of all concentrations of article/carrier mixtures on a monthly or quarterly basis.

When article/carrier mixtures are prepared by serial dilution of the highest concentration, FDA has suggested that it would be appropriate to perform an assay on the lowest concentration because this would confirm the accuracy of the dilution process. However, this is not a GLP requirement, and there is no prohibition on the analysis of any of the other concentrations. Analytical methods may not be sensitive enough for valid assays of the lowest concentration.

Although some laboratories do not use any article/carrier mixture until satisfactory analytical results are obtained from a concentration assay of the mixture, this is not a GLP requirement. The concentration assays provide periodic assurance that test systems are being exposed to the amounts and types of test and control articles that are specified in the protocol. Therefore, the results of the periodic concentration assays must be reviewed critically and promptly. Analytical results outside a pre-established acceptable range (as defined by laboratory SOPs) will require follow-up. Follow-up should attempt to determine the cause of poor analytical results (e.g., improper preparation of the article/carrier mixture, sample mix-up, poor analytical technique, equipment malfunction, etc.) Corrective action should then be provided as necessary. Usually analytical results in excess of 10% above or below expected values will require follow-up.

Tests to establish the stability and homogeneity of article/carrier mixtures as well as the periodic concentration analyses of the mixtures must be conducted in full compliance with the GLP regulations.

(b) (Reserved)

(c) Where any of the components of the test or control article carrier mixture has an expiration date, that date shall be clearly shown on the container. If more than one component has an expiration date, the earliest date shall be shown.

A reasonable interpretation of § 58.113(c) should not require expiration dating of containers of article/carrier mixtures when the mixtures will be used on the date of preparation unless a component of the mixture has an extremely short (e.g., less than 8 hours) period of stability. This section does not require that an expiration date appear on feeders which are filled with article/diet mixtures on the date the mixture is prepared and are presented to the test animals on that same day.

SUBPART G -- PROTOCOL FOR AND CONDUCT OF A NONCLINICAL LABORATORY STUDY

§ 58.120 Protocol

    (a)   Each study shall have an approved written protocol that clearly indicates the objectives and all methods for the conduct of the study. The protocol shall contain, as applicable, the following information:

The requirement to indicate "all methods for the conduct of the study" does not mean that all laboratory SOPs must be reiterated in the protocol. It is sufficient if the protocol indicates "what" will be done and "when" it will be done. Laboratory SOPs describe "how" each study activity is to be performed. If exceptions from SOPs will apply for the study, then those exceptions should be described in the protocol. FDA has indicated that the protocol should list the SOPs used in a particular study, but the author suggests that a simple stipulation in the protocol that "the study will be conducted in accordance with current standard operating procedures" is sufficient. Listing each SOP in the protocol could cause problems if SOP identifying numbers or titles change during the course of a study.

All of the following items, if relevant, must be included in the protocol:

    (1)   A descriptive title and statement of the purpose of the study.

    (2)   Identification of the test and control articles by name, chemical abstract number or code number.

    (3)   The name of the sponsor and the name and address of the testing facility at which the study is being conducted.

    (4)   The number, body weight range, sex, source of supply, species, strain, substrain, and age of the test system.

    (5)   The procedure for identification of the test system.

    (6)   A description of the experimental design, including the methods for the control of bias.

    (7)   A description and/or identification of the diet used in the study as well as solvents, emulsifiers and/or other materials

used to solubilize or suspend the test or control articles before mixing with the carrier. The description shall include specifications for acceptable levels of contaminants that are reasonably expected to be present in the dietary materials and are known to be capable of interfering with the purpose or conduct of the study if present at levels greater than established by the specifications.

If a laboratory conducts a blanket analysis for contaminants, the protocol can make reference to a description of those analyses in laboratory SOPs. Any additional analyses which are specific to the study should be described in the protocol.

(8) Each dosage level, expressed in milligrams per kilogram of body weight or other appropriate units, of the test or control article to be administered and the method and frequency of administration.

(9) The type and frequency of tests, analyses, and measurements to be made.

(10) The records to be maintained.

(11) The date of approval of the protocol by the sponsor and the dated signature of the study director.

(12) A statement of the proposed statistical methods to be used.

It is important to describe statistical methods in the protocol. This will avoid suspicions that statistical methods were selected after study data were available and that selection was based on a desired end result.

A protocol is required for each nonclinical laboratory study. Usually a single protocol will cover only one experiment with a single test article in a single type of test system. However, it is permissible to conduct several experiments using the same test article under a single, comprehensive protocol. It is also permissible to study several test articles concurrently using a single common procedure under one protocol.

The intent of § 58.120 is to provide all study personnel with clear directions as to the objectives of a study and all operations needed to fulfill those objectives. Therefore, even though not required by § 58.120, it is important that all personnel involved with a study have access to a copy of the protocol and all amendments. Such access best assures that study procedures will be done as and when intended.

(b)   All changes in or revisions of an approved protocol and the reasons therefor shall be documented, signed by the study director, dated, and maintained with the protocol.
(Collection of information requirements approved by the Office of Management and Budget under number 0910-0203)

Documentation of protocol changes or revisions and the reason for them is best accomplished by issuing formal protocol amendments, which must be dated and signed by the study director and should be attached to the front of all copies of the protocol. Such attachment immediately alerts study personnel to protocol changes and helps prevent study personnel from overlooking amendments which are "hidden" at the back of the protocol.

If deviations from a protocol are intended to be permanent, a protocol amendment should be issued to document the change. If a deviation from the protocol is an error, the deviation should be promptly corrected and should be documented in the study records and described in the final report.

To the extent possible, protocol amendments should be prospective, that is issued and distributed before the change is intended to occur. In some circumstances (e.g., an emergency decision to lower test article dose levels in a chronic study because of an unexpected toxic response to protocol-specified doses or a decision to collect additional tissue specimens where that decision is made on the basis of findings during the course of an autopsy) prospective distribution of a protocol amendment may not be possible. In such cases, a protocol amendment should be issued as soon as possible.

The question frequently arises as to what date should appear on a protocol amendment. The author is of the opinion that an effective date, whether prospective or retrospective, should be included. An effective date will alert personnel to the date when the amendment goes into effect and will also provide an historical record of the time period during which the amendment was in effect. It may also be helpful to include an issue date.

§ 58.130 Conduct of a Nonclinical Laboratory Study

(a) The nonclinical laboratory study shall be conducted in accordance with the protocol.

(b) The test systems shall be monitored in conformity with the protocol.

§§ 58.130 (a) and (b) should not be regarded as a strait-jacket which prevents scientifically justified changes in research as a study progresses. Any changes in the research can occur as long as they are properly documented in the form of protocol amendments. There is no limit on the number of amendments.

(c) Specimens shall be identified by test system, study, nature, and date of collection. This information shall be located on the specimen container or shall accompany the specimen in a manner that precludes error in the recording and storage of data.

The proper identification of specimens is of obvious importance to the validity of a study. Some types of specimens (e.g., paraffin blocks and microscopic slides), because of size or nature of the material, do not lend themselves to labeling for all the items listed in § 58.130(c). In such cases the use of an alternative identification (e.g., accession numbers) is acceptable as long as the alternate identification can be translated into the required information.

Failure to include the "nature" of the specimen will not, in some instances, be contrary to the intent of the regulations. For example, a microscopic slide which contains liver tissue need not have "liver" written on the slide since the end user, the diagnosing pathologist, will not need the label to identify the tissue as liver. On the other hand, sections of tumor from a multiple tumor-bearing animal should be clearly labeled to indicate from which tumor the sections were taken.

The phrase "shall accompany the specimen" need not be strictly interpreted in the case of archive material. For example, a specimen labeled with an accession number can be stored in the specimen archives while the document which translates the accession number into the additional label information is stored in a separate document archives. As long as both the specimen and the associated document are readily retrievable, the intent of the regulations is met.

(d)   Records of gross findings for a specimen from postmortem observations should be available to a pathologist when examining that specimen histopathologically.

To better ensure that the pathologist will be prompted to provide microscopic follow-up to all grossly observed lesions, it is important that information on gross findings be available to the diagnosing pathologist.

There may be occasions when study design requires that information on gross findings be withheld from the diagnosing pathologist (e.g., in the case of totally blinded slide reading). This is permissible, but FDA does not believe that "blinding" is a preferred practice in histopathologic evaluation.

(e)   All data generated during the conduct of a nonclinical laboratory study, except those that are generated by automated data collection systems, shall be recorded directly, promptly, and legibly in ink. All data entries shall be dated on the day of entry and signed or initialed by the person entering the data. Any change in entries shall be made so as not to obscure the original entry, shall indicate the reason for such change, and shall be dated and signed or identified at the time of the change. In automated data collection systems, the individual responsible for direct data input shall be identified at the time of data input. Any change in automated data entries shall be made so as not to obscure the original entry, shall indicate the reason for change, shall be dated, and the responsible individual shall be identified.
(Collection of information requirements approved by the Office of Management and Budget under number 0910-0203)

All data must be recorded promptly (defined by Webster as "immediately"). Hand-recorded data must be recorded in ink (to prevent improper erasures and corrections).

A signature (or initials) and date are not required for every individual piece of data. It is sufficient, for example, to provide one signature (or initials) and date for all data collected during a single data collection session. The purpose of the signature or initials is to provide accountability for the data.

The cGMP regulations (10) require certain activities (e.g., charge-in of components) to be performed by one individual and witnessed and verified by a second individual. There is no similar requirement in the GLP regulations, but some laboratories voluntarily elect to have certain critical operations (e.g., test article weighings) witnessed and verified by a second individual.

With the exception of automated data collection systems, all changes in data should be made by drawing a single line through the data being changed, recording the corrected or changed information and the date of change, and indicating a reason for the change. The person making the change should be identified by a signature or initials. The explanation of the change need not be elaborate. For example, "number transposition" or "entered in wrong column" can suffice as an explanation. Simply indicating "error" is seldom an adequate explanation. A coded system (e.g., number or letter) of recording the reasons for data changes is acceptable if the code is translated on the data form or in laboratory SOPs. The need to document reasons for changes in data must be constantly reinforced with study personnel.

Special rules apply in the case of automated data collection systems: the person responsible for data collection must be identified at the time of data input. Changes in automated data entries must be made in such a way that the original entry is saved, and the person responsible for making the change must be identified. The other requirements for data changes, recording the reason for the change and the date the change was made, also apply to automated data collection systems. The audit trail for changes in automated data entries may be recorded on paper or on computer.

SUBPARTS H-I -- (RESERVED)

SUBPART J -- RECORDS AND REPORTS

§ 58.185 Reporting of Nonclinical Laboratory Study Results

    (a)   A final report shall be prepared for each nonclinical laboratory study and shall include, but not necessarily be limited to, the following:

With the exception of the second sentence of item (7), all of the following topics must be addressed in the final report. Unlike § 58.120-(a), the words "as applicable" do not appear in § 58.185(a). Thus, for example, the report must address the issue of statistical analysis even if no statistical analysis was required or done.

    (1)   Name and address of the facility performing the study and the dates on which the study was initiated and completed.

FDA requires the name and address of the testing facility to appear in the report so that when the report is submitted in support of a research or marketing permit, the laboratory can be added to FDA's inventory of laboratories which are scheduled for GLP inspection. The name and address may also be used by FDA to establish the site for any directed audit of the report.

    (2)   Objectives and procedures stated in the approved protocol, including any changes in the original protocol.
    (3)   Statistical methods employed for analyzing the data.
    (4)   The test and control articles identified by name, chemical abstracts number or code number, strength, purity, and composition or other appropriate characteristics.
    (5)   Stability of the test and control articles under the conditions of administration.

The "stability...under the conditions of administration" will in most cases be the stability of the article/carrier mixtures determined under § 58.113(a)(2). If a drug is administered as a powder (e.g., by capsule), the stability of the bulk drug determined under § 58.105(b) will be reported.

(6) A description of the methods used.

(7) A description of the test system used. Where applicable, the final report shall include the number of animals used, sex, body weight range, source of supply, species, strain and substrain, age, and procedure used for identification.

(8) A description of the dosage, dosage regimen, route of administration, and duration.

(9) A description of all circumstances that may have affected the quality or integrity of the data.

Under § 58.33(c) the study director is responsible for documenting all circumstances that may affect the quality and integrity of the study. Such circumstances must be described in the final report.

(10) The name of the study director, the names of other scientists or professionals, and the names of all supervisory personnel, involved in the study.

Only names of study personnel need be listed in the report. Signatures required are those of the study director and those individuals described in § 58.185(a)(12). A laboratory is permitted some discretion in the listing of names. The names of technicians and animal-care workers need not be listed. The list of names is usually limited to senior scientific or supervisory staff.

(11) A description of the transformations, calculations, or operations performed on the data, a summary and analysis of the data, and a statement of the conclusions drawn from the analysis.

(12) The signed and dated reports of each of the individual scientists or other professionals involved in the study.

In the preamble (¶ 48a) to the 1987 GLP revisions (4), FDA rejected a request to modify § 58.185(a)(12) to permit combined reports signed by the principal scientists (e.g., clinical veterinarian, clinical pathologist, histopathologist, etc). FDA stated that each individual scientist involved in a study must be accountable for reporting data, information, and views within his or her designated area of responsibility and that combined reports would obscure the individual's accountability for accurate reporting.

Prior to publication of the 1987 GLP revisions, many laboratories prepared combined reports, and the author knows of no instance where the FDA rejected a study for failure to provide signed and dated reports from each of the scientists or other professionals involved in the study. For such laboratories it is probably advisable to reconsider prior policy on report preparation. The intent of the regulation (to provide account-ability) can be met with the format of a combined report, but with an indication on the signature page of the portion of the report prepared by each signatory.

It is customary to append the signed reports of consultants (e.g., consulting ophthalmologists, consulting pathologists, etc.) to the reports submitted to FDA.

(13)   The location where all specimens, raw data, and the final report are to be stored.

In most cases these materials will be stored in the archives of the testing facility, and the report will so indicate. However, in the case of contract safety testing, a sponsor will sometimes ask that raw data, documentation, and specimens be sent to the sponsor for storage in the sponsor's archives. In other cases a laboratory may store some or part of the archival material at an off-site location. In either case the final report should reference the actual storage site(s).

(14)   The statement prepared and signed by the quality assurance unit as described in § 58.35(b)(7).

(b)   The final report shall be signed and dated by the study director.

(c)   Corrections or additions to a final report shall be in the form of an amendment by the study director. The amendment shall clearly identify that part of the final report that is being added to or corrected and the reasons for the correction or addition, and shall be signed and dated by the person responsible.

A report becomes "final" when it is signed by the study director. Any changes in the report after it is signed by the study director must be in the form of an amendment which meets the requirements of § 58.185-(c). To avoid the necessity for many report amendments, the report should not be signed by the study director until it has been reviewed by the scientists involved in the study and has been audited by the quality assurance unit and after all changes and corrections occasioned by that review and audit have been made.

The purpose of § 58.185(c) is to guard against inappropriate or unwarranted changes being made in the report without the knowledge and concurrence of the study director.

§ 58.190 Storage and Retrieval of Records and Data

(a)   All raw data, documentation, protocols, final reports and specimens (except those specimens obtained from mutagenicity tests and wet specimens of blood, urine, feces, and biological fluids) generated as a result of a nonclinical laboratory study shall be retained.

(b)   There shall be archives for orderly storage and expedient retrieval of all raw data, documentation, protocols, specimens, and interim and final reports. Conditions of storage shall minimize deterioration of the documents or specimens in accordance with the requirements for the time period of their retention and the nature of the documents or specimens. A testing facility may contract with commercial archives to provide a repository for all material to be retained. Raw data and specimens may be retained elsewhere provided that the archives have specific reference to those other locations.

Any laboratory which conducts nonclinical laboratory studies must provide dedicated space for the storage of raw data, documentation, protocols, specimens, and interim and final reports from completed studies. The laboratory must have an orderly system for storing such material, and that system must provide an expedient method for retrieval of archived materials (for example, on the request of an FDA inspector).

Storage conditions (e.g., temperature, humidity, etc.) in the archives should be reasonably related to the nature of the stored documents, specimens, and samples. For example, wet tissues and paraffin blocks should be protected against extremes of high temperature; paper documents should not be subjected to long periods of high humidity; reserve samples of test and control articles should be stored in accordance with label requirements. FDA has indicated that "heroic" measures need not be taken to preserve materials in the archives, but storage conditions which foster accelerated deterioration should be avoided. Storage conditions should be monitored so that deviations from proper storage conditions can be promptly rectified.

If an off-site area is used to house the archives, whether owned or rented by the testing facility or operated by a commercial archival service, the on-site archives must contain specific reference to the materials that are stored off-site and the location of the alternate storage site(s).

(c) An individual shall be identified as responsible for the archives.

Similar to the requirements for a study director, a nonclinical testing laboratory must designate a single individual to be responsible for the archives. It is permissible to designate an alternate archivist to serve in the absence of the designated archivist.

(d) Only authorized personnel shall enter the archives.

Laboratory SOPs should define the personnel who may enter the archives. This need not be a list of names of individuals, but should provide adequate guidance to archive personnel as to who may enter the archives. Many laboratories allow only archive personnel to enter the archives but allow authorized personnel to check out archive material. If materials are removed from the archives for any reason, a record should be kept of what is removed and by whom. Follow-up should be provided by archive personnel to ensure prompt return of materials to the archives.

(e) Material retained or referred to in the archives shall be indexed to permit expedient retrieval.
(Collection of information requirements approved by the Office of Management and Budget under number 0910-0203)

Any indexing system for material in the archives is acceptable as long as the system permits rapid retrieval of archived materials.

§ 58.195   Retention of Records

(a) Record retention requirements set forth in this section do not supersede the record retention requirements of any other regulations in this chapter.

If the record retention requirements of § 58.195 are inconsistent with those of any other part of 21 CFR, the other parts of 21 CFR will take precedence.

(b) Except as provided in paragraph (c) of this section, documentation records, raw data and specimens pertaining to a nonclinical laboratory study and required to be made by this part shall be retained in the archive(s) for whichever of the following periods is shortest:
(1) A period of at least 2 years following the date on which an application for a research or marketing permit, in support of which the results of the nonclinical laboratory study were submitted, is approved by the Food and Drug Administration. This requirement does not apply to studies supporting

investigational new drug applications (IND's) or applications
for investigational device exemptions (IDE's), records of
which shall be governed by the provisions of paragraph
(b)(2) of this section.

(2)   A period of at least 5 years following the date on which the
results of the nonclinical laboratory study are submitted to
the Food and Drug Administration in support of an application for a research or marketing permit.

(3)   In other situations (e.g., where the nonclinical laboratory
study does not result in the submission of the study in
support of an application for a research or marketing permit),
a period of at least 2 years following the date on which the
study is completed, terminated, or discontinued.

Records, raw data, and specimens from a nonclinical laboratory
study must be retained for whichever of the three time periods indicated
above is shortest. An exception is made for those nonclinical laboratory
studies which support an application for an IND or an IDE for which
records must be retained for a minimum of 5 years after the results of
those studies are submitted to FDA.

Most companies take a more conservative approach and retain
documents, microscopic slides, and paraffin blocks indefinitely. Materials
such as wet tissues which take up more storage space are generally the
first materials to be discarded. Paper documents may be discarded at any
time if they have been converted to microfilm or microfiche.

(c)   Wet specimens (except those specimens obtained from
mutagenicity tests and wet specimens of blood, urine, feces,
and biological fluids), samples of test or control articles, and
specially prepared material which are relatively fragile and
differ markedly in stability and quality during storage, shall
be retained only as long as the quality of the preparation
affords evaluation. In no case shall retention be required for
longer periods than those set forth in paragraphs (a) and (b)
of this section.

If a laboratory elects to discard fragile materials before the expiration of the applicable time period of § 58.195(b), the date of discard and the justification for discard should be recorded, and the documentation should be retained in the archives.

Examples of "specially prepared material" were listed in the GLP regulations prior to the 1987 revisions. These included histochemical, electron microscopic, blood mounts, and teratological preparations. These examples are illustrative and not comprehensive.

(d) The master schedule sheet, copies of protocols, and records of quality assurance inspections, as required by § 58.35(c) shall be maintained by the quality assurance unit as an easily accessible system of records for the period of time specified in paragraphs (a) and (b) of this section.

The records and documents required to be maintained by the quality assurance unit are also subject to the record retention requirements of § 58.195(b). FDA spokesmen have stated on occasion that these quality assurance records should be stored in the archives described in § 58.190-(b). This is an option that can be considered by the quality assurance unit, but there is no stipulated requirement in the GLP regulations for such storage. In fact, it could be argued that the requirement of § 58.35(a) for the quality assurance function to be independent of nonclinical laboratory study personnel militates against storage of quality assurance records in the archives.

(e) Summaries of training and experience and job descriptions required to be maintained by § 58.29(b) may be retained along with all other testing facility employment records for the length of time specified in paragraphs (a) and (b) of this section.

Rather than storing summaries of training and experience and job descriptions in the GLP archives, a laboratory may elect to store such records together with other employment records (e.g., in the Personnel Department). If such alternative storage of these records is elected, care should be taken that personnel responsible for the alternate records storage are aware of GLP record retention requirements. Before electing such alternate storage, a system should be established to preserve the confidentiality of the personnel records (other than summaries of training and experience and job descriptions) at the time of FDA or quality assurance inspections.

> (f) Records and reports of the maintenance and calibration and inspection of equipment, as required by § 58.63(b) and (c), shall be retained for the length of time specified in paragraph (b) of this section.

Records of the maintenance, calibration, and inspection of equipment are also subject to the record retention requirements of the regulations. Often a facility has its own metrology group or contracts with an outside group to handle maintenance and calibration of equipment. In such cases the records of these activities may include records for equipment which are used in both GLP and non-GLP studies, and may be stored centrally within an organization and outside the GLP archives. This is not contrary to GLP requirements as long as the regular GLP archives makes reference to the alternate storage place and as long as the alternate storage meets the GLP requirements for secure and orderly storage, expedient retrieval of records, limited access to the records storage area, and responsibility for storage under the control of a single individual.

> (g) Records required by this part may be retained either as original records or as true copies such as photocopies, microfilm, microfiche, or other accurate reproductions of the original records.

The 1987 GLP revisions added § 58.195(g) to re-emphasize long-standing FDA policy that a laboratory may retain either original records or accurate reproductions of them. It should be noted that magnetic media may qualify as either original records or accurate reproductions of same.

> (h) If a facility conducting nonclinical testing goes out of business, all raw data, documentation, and other material

specified in this section shall be transferred to the archives of the sponsor of the study. The Food and Drug Administration shall be notified in writing of such a transfer.

A laboratory going out of business is often a sudden and unplanned event. Under such circumstances, personnel from the lab ceasing operations may not show proper concern for complying with regulatory requirements. Therefore, the party with the greatest stake in preserving the records of a study, namely the sponsor, may have to assume responsibility for preservation and transfer of the records to the sponsor's location and for notifying FDA of the transfer.

There was an instance where a laboratory which had conducted a number of studies for EPA regulatory purposes went out of business, and the records relating to studies the laboratory had conducted were lost. EPA in this case required many of the studies to be repeated. The lesson to be learned from this experience is that a sponsor should be very careful in the selection of contract facilities and should periodically check with the contract lab to ensure that the laboratory continues to operate and that study records continue to be maintained. Some sponsors obtain the specimens and/or originals or copies of all raw data for contracted studies for storage in their own archives to protect against the loss of raw data at the contract laboratory.

SUBPART K - DISQUALIFICATION OF TESTING FACILITIES

§ 58.200 Purpose

(a) The purposes of disqualification are: (1) To permit the exclusion from consideration of completed studies that were conducted by a testing facility which has failed to comply with the requirements of the good laboratory practice regulations until it can be adequately demonstrated that such noncompliance did not occur during, or did not affect the validity or acceptability of data generated by, a particular study; and (2) to exclude from consideration all studies completed after the date of disqualification until the facility can satisfy the Commissioner that it will conduct studies in compliance with such regulations.

Disqualification is the most severe penalty that FDA can apply for failure to comply with GLP requirements. If a laboratory is disqualified, the completed or future studies conducted by that laboratory may not be accepted by FDA in support of an application for a research or marketing permit. It is even possible for prior FDA approval of a marketed product to be withdrawn if that approval was based in part on the study or studies conducted by a disqualified laboratory.

(b)   The determination that a nonclinical laboratory study may not be considered in support of an application for a research or marketing permit does not, however, relieve the applicant for such a permit of any obligation under any other applicable regulation to submit the results of the study to the Food and Drug Administration.

If a sponsor is actively pursuing a research or marketing permit for a test article and if a disqualified laboratory has conducted a nonclinical laboratory study on that test article, the sponsor is still obligated to submit the results of such a study to the FDA. As indicated in paragraph (a), FDA will not consider the results of that study in support of the research or marketing permit, but FDA may use the results of the study in reaching a conclusion that the research or marketing permit should not be approved. Thus the results of a study conducted by a disqualified laboratory can never work to the sponsor's advantage but may work to the sponsor's disadvantage.

§ 58.202 Grounds for Disqualification

The Commissioner may disqualify a testing facility upon finding all of the following:

(a)   The testing facility failed to comply with one or more of the regulations set forth in this part (or any other regulations regarding such facilities in this chapter);

(b)   The noncompliance adversely affected the validity of the nonclinical laboratory studies; and

(c)   Other lesser regulatory actions (e.g., warnings or rejection of individual studies) have not been or will probably not be adequate to achieve compliance with the good laboratory practice regulations.

It is important to note that the FDA must find all 3 conditions, as indicated in paragraphs (a), (b), and (c) above, before it can disqualify a laboratory.

Since the effective date of the GLP regulations, no laboratory has been disqualified. FDA has, however, issued many warnings and has rejected individual studies for reasons of GLP noncompliance.

There are instances where laboratories have gone out of business "voluntarily" because they lack the desire or ability to comply with GLP requirements.

With the exception of § 58.217, the balance of Subpart K describes the legal and administrative procedures which govern the disqualification process. The remaining sections of Subpart K are reprinted for the sake of completeness but will not be commented on with the exception of § 58.217.

§ 58.204 Notice of and Opportunity for Hearing on Proposed Disqualification

(a)   Whenever the Commissioner has information indicating that grounds exist under § 58.202 which in his opinion justify disqualification of a testing facility, he may issue to the testing facility a written notice proposing that the facility be disqualified.

(b)   A hearing on the disqualification shall be conducted in accordance with the requirements for a regulatory hearing set forth in Part 16.

§ 58.206 Final Order on Disqualification

(a) If the Commissioner, after the regulatory hearing, or after the
     time for requesting a hearing expires without a request being
     made, upon an evaluation of the administrative record of the
     disqualification proceeding, makes the findings required in
     § 58.202, he shall issue a final order disqualifying the
     facility. Such order shall include a statement of the basis for
     that determination. Upon issuing a final order, the Commis-
     sioner shall notify (with a copy of the order) the testing
     facility of the action.

(b) If the Commissioner, after a regulatory hearing or after the
     time for requesting a hearing expires without a request being
     made, upon an evaluation of the administrative record of the
     disqualification proceeding, does not make the findings
     required in § 58.202, he shall issue a final order terminating
     the disqualification proceeding. Such order shall include a
     statement of the basis for that determination. Upon issuing
     a final order the Commissioner shall notify the testing facility
     and provide a copy of the order.

§ 58.210 Actions upon Disqualification

(a) Once a testing facility has been disqualified, each application
     for a research or marketing permit, whether approved or not,
     containing or relying upon any nonclinical laboratory study
     conducted by the disqualified testing facility may be exam-
     ined to determine whether such study was or would be
     essential to a decision. If it is determined that a study was
     or would be essential, the Food and Drug Administration
     shall also determine whether the study is acceptable, notwith-
     standing the disqualification of the facility. Any study done
     by a testing facility before or after disqualification may be
     presumed to be unacceptable, and the person relying on the
     study may be required to establish that the study was not
     affected by the circumstances that led to the disqualification,

e.g., by submitting validating information. If the study is then determined to be unacceptable, such data such (sic) be eliminated from consideration in support of the application; and such elimination may serve as new information justifying the termination or withdrawal of approval of the application.

(b) No nonclinical laboratory study begun by a testing facility after the date of the facility's disqualification shall be considered in support of any application for a research or marketing permit, unless the facility has been reinstated under § 58.219. The determination that a study may not be considered in support of an application for a research or marketing permit does not, however, relieve the applicant for such a permit of any obligation under any other applicable regulation to submit the results of the study to the Food and Drug Administration.

§ 58.213 Public Disclosure of Information Regarding Disqualification

(a) Upon issuance of a final order disqualifying a testing facility under § 58.206(a), the Commissioner may notify all or any interested persons. Such notice may be given at the discretion of the Commissioner whenever he believes that such disclosure would further the public interest or would promote compliance with the good laboratory practice regulations set forth in this part. Such notice, if given, shall include a copy of the final order issued under § 58.206(a) and shall state that the disqualification constitutes a determination by the Food and Drug Administration that nonclinical laboratory studies performed by the facility will not be considered by the Food and Drug Administration in support of any application for a research or marketing permit. If such notice is sent to another Federal Government agency, the Food and Drug Administration will recommend that the agency also consider whether or not it should accept nonclinical laboratory studies performed by the testing facility. If such notice is sent to any other person, it shall state that it is given because of the relationship between the testing facility and the person being notified and that the Food and Drug Administration is not advising or recommending that any action be taken by the person notified.

(b)   A determination that a testing facility has been disqualified and the administrative record regarding such determination are disclosable to the public under Part 20.

§ 58.215   Alternative or Additional Actions to Disqualification

(a)   Disqualification of a testing facility under this subpart is independent of, and neither in lieu of nor a precondition to, other proceedings or actions authorized by the act. The Food and Drug Administration may, at any time, institute against a testing facility and/or against the sponsor of a nonclinical laboratory study that has been submitted to the Food and Drug Administration any appropriate judicial proceedings (civil or criminal) and any other appropriate regulatory action, in addition to or in lieu of, and prior to, simultaneously with, or subsequent to, disqualification. The Food and Drug Administration may also refer the matter to another Federal, State, or local government law enforcement or regulatory agency for such action as that agency deems appropriate.

(b)   The Food and Drug Administration may refuse to consider any particular nonclinical laboratory study in support of an application for a research or marketing permit, if it finds that the study was not conducted in accordance with the good laboratory practice regulations set forth in this part, without disqualifying the testing facility that conducted the study or undertaking other regulatory action.

§ 58.217   Suspension or Termination of a Testing Facility by a Sponsor

Termination of a testing facility by a sponsor is independent of, and neither in lieu of nor a precondition to, proceedings or actions authorized by this subpart. If a sponsor terminates or suspends a testing facility from further participation in a nonclinical laboratory study that is being conducted as part of any application for a research or marketing permit that has been submitted to any Center of the Food and Drug Administration (whether approved or not), it shall notify that Center in writing within 15 working days of the action; the notice shall include a statement of the reasons for such action. Suspension or termination of a testing facility by a sponsor does not relieve it of any obligation under any other applicable regulation to submit the results of the study to the Food and Drug Administration.

Under the provisions of § 58.217, if a sponsor, for any reason, terminates or suspends a testing facility from further participation in a nonclinical laboratory study and if the test article in that study is the subject of any application to FDA for a research or marketing permit, then the sponsor must notify FDA, in writing and within 15 working days, of the termination or suspension. The notice to FDA must also include the reason for the termination or suspension.

§ 58.219   Reinstatement of a Disqualified Testing Facility

A testing facility that has been disqualified may be reinstated as an acceptable source of nonclinical laboratory studies to be submitted to the Food and Drug Administration if the Commissioner determines, upon an evaluation of the submission of the testing facility, that the facility can adequately assure that it will conduct future nonclinical laboratory studies in compliance with the good laboratory practice regulations set forth in this part and, if any studies are currently being conducted, that the quality and integrity of such studies have not been seriously compromised. A disqualified testing facility that wishes to be so reinstated shall present in writing to the Commissioner reasons why it believes it should be reinstated and a detailed description of the corrective actions it has taken or intends to take to assure that the acts or omissions which led to its disqualification will not recur. The Commissioner may condition reinstatement upon the testing facility being found in compliance with the good laboratory practice regulations upon an inspection. If a testing facility is reinstated, the Commissioner shall so notify the testing facility and all organizations and persons who were notified, under § 58.213 of the disqualification of the testing facility. A determination that a testing facility has been reinstated is disclosable to the public under Part 20.

CONFORMING AMENDMENTS

At the time of *Federal Register* publication of final GLP regulations, FDA also made amendments to a multitude of other sections of 21 CFR. These so-called conforming amendments all require that a statement be included with respect to each nonclinical laboratory study which is submitted to FDA in support of an application for a research or marketing permit. The conforming amendment statement can be in either of two forms.

If the study was conducted in full compliance with GLP requirements, the conforming amendments statement will so indicate. If not, then the conforming amendments statement must contain a brief statement of the reason for the noncompliance.

FDA has required a conforming amendments statement for all nonclinical laboratory studies submitted to FDA after June 20, 1979, the effective date of the GLP regulations. Thus, a conforming amendments statement was required for studies completed prior to June 20, 1979, if the results of the studies were submitted to FDA after that date.

When several nonclinical laboratory studies are contained in a single submission to FDA, a single conforming amendments statement may be included with the submission, or the sponsor may elect to prepare individual statements for each study.

Preparation of the conforming amendments statement is the responsibility of the sponsor of the study even if the study was conducted by someone other than the sponsor. This is consistent with FDA's view that ultimate responsibility for a study rests with the sponsor. In the case of contracted studies, the sponsor should ask the contractor to supply the information necessary to enable the sponsor to prepare a proper conforming amendments statement. FDA has not specified who should sign the conforming amendments statement. Generally it will be the same individual who signs the official application for a research or marketing permit. However, if a statement is included with the report of each study submitted to the FDA, the statement may be signed by the study director, by laboratory management, by quality assurance personnel, or by a combination of those individuals.

FDA has indicated that the conforming amendments statement can be brief for studies, such as preliminary exploratory studies and studies conducted prior to the effective date of the GLP regulations, which are exempt from GLP requirements. In such cases the statement need only indicate the GLP-exempt status of the studies.

GLP deviations that were of a continuing nature throughout the course of a study will require a conforming amendments statement of the reason for the noncompliance. One-time deviations from GLP requirements should be documented in study records and should be described in the final report but will not require a conforming amendments statement of the reason for the noncompliance.

Care should be taken in the preparation of the conforming amendments statements. While failure to comply with GLPs is only subject to administrative sanctions (e.g., disallowance of a study or disqualification of a testing facility), knowingly submitting a false statement to the FDA is a criminal offense punishable by fine and/or imprisonment.

REFERENCES

1. *Federal Register.* 41: 51206 (1976).

2. *Federal Register.* 43: 59986 (1978), corrected *Federal Register.* 44: 17657 (1979).

3. *Federal Register.* 45: 2486 (1980).

4. *Federal Register.*   52: 33768 (1987), corrected *Federal Register.* 52: 36863 (1987).

5. *Federal Register.*   54: 9038 (1989).

6. *Federal Register.*   54: 15923 (1989).

7. *Federal Register.*   56: 32087 (1991).

8. 21 C.F.R. Part 58, Good Laboratory Practice for Nonclinical Laboratory Studies.

9. 21 U.S.C. § 371(a).

10. 21 C.F.R. Part 211, Current Good Manufacturing Practice for Finished Pharmaceuticals.

11. Food and Drug Administration, Good Laboratory Practice Regulations Management Briefings-Post Conference Report.

12. Food and Drug Administration, Dockets Management Branch [HFA-305] (available under Freedom of Information).

13. S. H. Willig, M. M. Tuckerman, and W. S. Hitchings IV, *Good Manufacturing Practices for Pharmaceuticals: A Plan for Total Quality Control*, 2nd Ed., Revised and Expanded, Marcel Dekker, Inc. (1982), p. 201.

14. S. H. Willig, and J. R. Stoker, *Good Manufacturing Practices for Pharmaceuticals: A Plan for Total Quality Control*, 3rd Ed., Marcel Dekker, Inc. (1991), p. 187.

15. 9 C.F.R. Subchapter A--Animal Welfare

16. G. O. Allen, A. F. Hirsch, and H. Leidy, Application of military STD 105D sampling plans to research report audits to meet GLP requirements," *Drug Information Journal* 14:65 (1980).

17. R. M. Siconolfi, Statistical approaches to auditing, in *Managing Conduct and Data Quality of Toxicology Studies*, Princeton Scientific Publishing Co., Inc. (1986), p. 223.

18. U.S. Department of Defense, MIL-C-45662A, 9 February 1962.

19. S. Z. Diamond, *Preparing Administrative Manuals*, AMACOM (1981).

20. P. Lepore, Presentation at PRIM&R's March 19-20, 1992 *The Animal Care Committee and Enforced Self-Regulation: Making it Work at Your Institution* conference in Boston.

21. U.S. Department of Health and Human Services, *Guide for the Care and Use of Laboratory Animals*, NIH Publication No. 85-23 (1985).

Chapter 3

EPA AND FOREIGN COUNTRY GLP REGULATIONS

Carl R. Morris, Ph.D.

International Chemical Consultants U.S.A., Inc., Alexandria, Virginia

Frederick G. Snyder

Astra/Merck Group of Merck & Co., Inc., Wayne, Pennsylvania

INTRODUCTION

During the late 1960s and early 1970s, many nations became increasingly concerned about the quality of their environments and the potential impacts of industrial chemicals on their lifestyles. The need to assess the balance between higher living standards and the problems associated with potential increased exposures to new chemical products caused by increased industrialization became a high priority for many countries. This concern prompted many countries to enact new environmental legislation or administrative procedures to address these issues. The lack of data for assessing the potential risks associated with these various chemical exposures resulted in the development of both voluntary and required testing programs to be conducted by private industry and by various governmental entities. As the results of these new testing initiatives began to enter the public arena through various environmental regulatory agencies, questions arose concerning the quality and integrity of the submitted data. In general, regulatory investigators observed that both sponsors and testing laboratories needed to increase their surveillance of the conduct of studies and recording of data by laboratory personnel.

When it was learned through regulatory facility inspections and audits of original study data that a few laboratories had submitted incomplete reports to government agencies and the documentation that supported these studies was often poor and, in some cases, could not be found, the regulatory agencies were left with the impression that these studies may be of questionable value for regulatory purposes. In addition, inspectors observed that many studies were poorly managed and conducted and, in some instances, there were questions as to the qualifications of personnel conducting these studies. These conditions not only pointed to the issues of data quality and integrity but also to the need for better systems to store and retrieve data. In addition, there was the impression that study sponsors did not adequately monitor their studies. These concerns resulted in the establishment of good laboratory management procedures [i.e., Good Laboratory Practice Regulations, (GLPs)] requiring laboratories to establish management structures and systems, including "quality assurance units" (QAUs) that would be independent of the actual conduct of studies. These QAUs or "third parties" serve as independent inspectors and auditors of studies to assure compliance with the approved protocols and the laboratories' standard operating procedures (SOPs). QAU findings are submitted to designated study directors and to upper-level management of the laboratory.

A growing awareness of the potential problems in instituting different testing requirements and different GLPs, made it clear that a major international effort was needed to standardize these various requirements in order to avoid potential non-tariff barriers to trade. Since 1977, the United States and other members of the 24-nation Organization for Economic Cooperation and Development (OECD) have been involved in extensive international consultations in efforts to bring industrial and environmental chemical programs into harmony.

The purpose of this organization is to help member countries promote sound economic growth, employment, and improve their respective standards of living. In addition, OECD's purpose is also to promote sound and harmonious development of the world economy with particular assistance for developing countries. As part of this philosophy, OECD established several expert groups, including one to address the issues of writing and implementing international guidelines for GLP (1). A series of expert group meetings was held between 1978 and 1981, at which several international testing guidelines were presented to the participating countries for review, comment, and approval.

Through the international OECD Expert Group on GLP, a major effort was directed toward the development of international guidelines for Good Laboratory Practice (GLP) (2). The principal objective of these guidelines was to assure, to the extent practicable under the laws of the OECD member countries, that data developed to meet one country's requirements would be acceptable to other countries. The United States and other member countries of the OECD placed a high priority on these activities because of the benefits for international chemical trade and for more effective health and environmental protection. There was strong endorsement of the work of the OECD Expert Group on GLP at meetings of high-level national regulatory officials in May 1980 and in November 1982. In May 1981, OECD member countries adopted a formal decision on the mutual acceptance of data that, to the extent practicable under the laws of OECD member countries, binds member countries to accept, for assessment purposes, data generated according to the OECD Test Guidelines and the OECD Principles of Good Laboratory Practice.

In addition to the development of the OECD Principles of GLP, the OECD Expert Group on GLP was given the responsibility for developing two additional guidance documents: one for the implementation of OECD Principles of GLP (2) and one to serve as OECD Guidelines for National GLP Inspections and Study Audits (2).

The implementation document encourages member countries to adopt the OECD Principles of GLP into their legislative and administrative frameworks, calls for documentation of compliance programs by national authorities, and requires declaration by each laboratory that the studies were conducted in accordance with the OECD Principles of GLP or with national regulations or equivalents conforming to these principles. The guideline recommends that national compliance programs utilize laboratory inspections and study audits as principal mechanisms for monitoring compliance with the Principles of GLP. It further recommends that national authorities utilize properly trained personnel who are competent to assess laboratory compliance with the principles and to administer the GLP compliance programs. The implementation document also advocates the inclusion in each national GLP compliance program actions to deal with noncompliance and deficiencies that may be taken by the national authorities.

The following sections examine the current status of several countries' efforts to meet these implementation objectives.

U.S. ENVIRONMENTAL PROTECTION AGENCY GLP IMPLEMENTATION

The U.S. Environmental Protection Agency (EPA) and the U.S. Food and Drug Administration (FDA) have been involved in numerous GLP-related activities since the early 1970s. Table 1 on page 116 presents various GLP milestones for the activities of these two regulatory agencies in implementing GLP Regulations and GLP compliance programs as well as their involvement in the international GLP arena.

EPA has proposed and finalized GLP Regulations for both the Toxic Substances Control Act (TSCA) and the Federal Insecticide, Fungicide, and Rodenticide Act (FIFRA). These efforts are intended to be consistent with each other except for differences required by their respective statutory authorities. In addition, the language of these regulations has been harmonized to the extent possible with the language of the GLP Regulations promulgated by the U.S. Food and Drug Administration (FDA).

Although the EPA's final GLP Regulations do not exactly reflect the Organization for Economic Cooperation and Development's (OECD) GLP guidance document, the Agency has stated that the OECD GLP principles are embodied in the final EPA GLP Regulations.

The scope of the existing GLP Regulations affects all types of health and environmental testing, including environmental fate testing and epidemiology studies. The 1989 amendments to these regulations include ecological effects, residue chemistry, efficacy testing, field studies, and epidemiology studies under FIFRA and field and epidemiology studies under TSCA.

These GLP Regulations provide guidance for ongoing industry testing when that testing is expected to be presented to the Agency for registration or health and safety assessment purposes. They also serve as one of the factors for EPA's decisions relating to required industry testing under Section 4 of TSCA. The GLP Regulations are included as a condition for industry testing under negotiated consent testing agreements wherein EPA and industry work together to further refine testing needs for priority chemicals identified by the Interagency Testing Committee or through Agency review and prioritization mechanisms. In addition, all industry testing under TSCA Section 4 Rule Making requires that the provisions in these GLP Regulations be used in the development of the required test data.

EPA GLP Compliance Programs

In 1978, EPA entered into an Interagency Agreement with FDA that formalized the agencies' cooperative efforts to establish a coordinated U.S. approach to quality assurance and GLP implementation. Under the terms of this agreement, FDA provides inspectional support for EPA programs that involve laboratories conducting health effects studies intended for submission to EPA. This approach permits FDA inspectors to carry out targeted inspections and/or study audits of studies for both FDA and EPA with only one inspection visit. This results in more efficient use of government resources and minimizes the use of testing laboratory resources in responding to regulatory needs. EPA's enforcement staff conducts inspections and study audits for environmental studies, including chemical fate testing for TSCA and FIFRA compliance.

In late 1993, the responsibilities for GLP Compliance Monitoring were moved from the Office of Prevention, Pesticides and Toxic Substances (OPPTS) to the Office of Compliance (OC) within the Office of Enforcement and Compliance Assurance (OECA). The OPTS Office of Compliance Monitoring staff that had been responsible for establishing and managing the Agency's GLP implementation and compliance program was moved to this new program office along with their continuing compliance monitoring responsibilities (see Figure 1). The Staff of the Agricultural and Ecosystems Division of the Office of Compliance is charged with the responsibility to assure the quality and integrity of data submitted to the Agency under TSCA and FIFRA. This is accomplished by implementing and utilizing the TSCA and FIFRA Good Laboratory Practice Regulations, including administrative

Table 1
Good Laboratory Practice (GLP) Milestones

1976: FDA Proposal (Good Laboratory Practice for Non-Clinical Laboratory Studies) November 19, 1976; *Federal Register;* 41:51206.

1978: FDA Final Rule (Good Laboratory Practice for Non-Clinical Laboratory Studies) December 22, 1987; *Federal Register*, 43:59986.

1979: EPA Proposal (TSCA: Health Effects) May 9, 1979; *Federal Register*, 44:27362.

1980: EPA Proposal (FIFRA: Health Effects) April 18, 1980; *Federal Register*, 45:26373. EPA Proposal (TSCA: Environmental Effects & Chemical Fate) November 21, 1980; *Federal Register*,44:77357.

1981: Principles of Good Laboratory Practice. Organization for Economic Cooperation and Development (OECD), Paris, France 1981. Guidelines for National GLP Inspections and Study Audits, Organization for Economic Cooperation and Development (OECD), Paris, France, 1981.

1983: EPA Final Rule (TSCA: Health & Environmental Effects & Fate of Chemical Substances & Mixtures) November 29, 1983; *Federal Register*, 48:53922. EPA Final Rule (FIFRA: Health Effects); November 29, 1983; *Federal Register*, 48:53946.

1984: FDA Amendment Proposal (Good Laboratory Practice for Non-Clinical Laboratory Studies); October 19, 1984; *Federal Register*, 49:43530.

1987: FDA Amendment Final Rule (Good Laboratory Practice for Non-Clinical Laboratory Studies); September 4, 1987; *Federal Register*, 52:33768. EPA Proposal (FIFRA: Ecological Effects, Chemical Fate, Residue Chemistry, & Product Performance, i.e., Efficacy Testing); December 28, 1987; *Federal Register*, 52:48920. EPA Amendment Proposal (FIFRA: Health Effects); December 28, 1987; Federal Register, 52:48920. EPA Amendment Proposal (TSCA: Health & Environmental Effects, Chemical Fate, & Field Studies); December 28, 1987; *Federal Register*, 52:48933.

1989: EPA Amendment Final Rule (FIFRA: Health Effects); August 17, 1989; *Federal Register*, 54: 34052. EPA Final Rule (FIFRA: Ecological Effects, Chemical Fate, Residue Chemistry, & Product Performance, i.e., Efficacy Testing); August 17, 1989; *Federal Register*, 54: 34052. EPA Amendment Final Rule (TSCA: Health & Environmental Effects, Chemical Fate, & Field Studies); August 17, 1989; *Federal Register*, 54:34034.

and compliance procedures such as on-site inspections and study audits of testing facilities and internal Agency audits of submitted data. Program office scientists participate in inspections and study audits of laboratories on a case-by-case basis.

Figure 1
EPA GLP Organizational Compliance Program

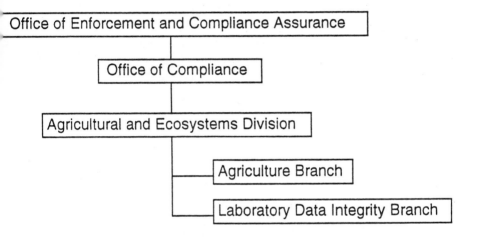

EPA GLP ENFORCEMENT RESPONSE POLICY

On April 9, 1985, the EPA Office of Compliance Monitoring issued a final Enforcement Response Policy (ERP) for the Toxic Substances Control Act (TSCA) Good Laboratory Practice (GLP) Regulations which were published on November 29, 1983 (48 FR 53922). The policy document cites levels of action that the Agency will take when the GLPs are not followed. The most common responses to violations of TSCA GLP Regulations are Notices of Noncompliance and civil administrative penalties. Notices of Noncompliance generally involve minor or technical violations that do not, either separately or collectively, have an impact upon the Agency's ability to evaluate chemical substances or mixtures.

The submission of "false" GLP Compliance Statements by the study director or sponsor has been the most common types of violation that has been reported and prosecuted. EPA utilizes civil administrative penalties for most other violations. Under some extreme situations, EPA has sought criminal sanctions for the most serious violations (e.g., fraudulent data).

If studies that are submitted under Section 5 of TSCA are not conducted in accordance with GLP requirements, the Agency may elect to consider the data insufficient or unreliable to evaluate the health and environment effects and fate of the chemical. These situations may also result in the issuance of a Notice of Noncompliance.

On September 30, 1991, EPA's Office of Compliance Monitoring also issued an Enforcement Response Policy (ERP) for the Federal Insecticide, Fungicide, and Rodenticide Act (FIFRA) Good Laboratory Practice (GLP) Regulations. The levels of enforcement action for violations of the FIFRA GLP Regulations include Notices of Warning, administrative civil penalties, and criminal proceedings. The Agency may also announce their actions through press releases. This latter approach places the testing facility under public scrutiny and may impact their relationships with current or potential sponsors.

INTERNATIONAL GLP IMPLEMENTATIONS

The Principles of Good Laboratory Practices (GLPs) imposed by the government of the United States via the Food and Drug Administration (FDA) and the Environmental Protection Agency (EPA) have been discussed extensively because they generated intense levels of activity and put significant pressure on domestic and foreign firms that wished to submit safety data in support of product registration and licensing in the United States. The progress that has been made in implementing these principles on a worldwide basis has been significant.

The member countries of both OECD and the European Union (EU) [formerly the European Communities] have played an integral part in the implementation of the GLP principles, particularly by assisting the various government authorities in establishing Good Laboratory Practices as a universal standard. Directives of the OECD in the late 1970s and early 1980s and, more recently, of the EU have done much to assure that the managerial principles of the GLPs are universally applied as unified, workable concepts. The countries (3) that presently have GLPs in place, or are in the process of adopting the OECD GLP principles in some fashion, are discussed below.

Except in the case of Japan, a general overview of programs in place or in the process of being adopted is presented. Because of the intense activity level in Japan and its potential impact on the international regulated community, a more detailed discussion is presented.

Canada

Within the Canadian government, there are two ministries that impact on the implementation of GLP principles. The first is the Health Protection Branch (HPB) under the Ministry of Health and Welfare of Canada, which is responsible for the establishment and implementation of GLP standards relative to drugs, food additives, etc. The second ministry, Environment Canada, deals with environmental issues and, therefore, is responsible for industrial chemicals.

Although Canada is generally influenced by action taken by its neighbor, the United States, for reasons that are unclear, the Canadian government has been reluctant to move forward in promulgating GLP Regulations. The rationale most often given is that "domestically" they have very few toxicology laboratories, and, therefore, GLPs are a low national priority.

Belgium

As a member of the EU, they have established a GLP program in the Ministry of Public Health and the Environment with the Institute of Hygiene and Epidemiology providing the Inspectorate to assure compliance with the EC Directive of December 18, 1989; effective July 1, 1990.

Denmark

Initially (1988), the Danish Ministry of the Environment published GLP regulations for industrial chemicals testing programs. This responsibility was later transferred to the National Agency of Industry and Trade which re-promulgated the GLP regulations on March 19, 1991. At present, the Danish government, through the Danish Accreditation Scheme (formerly the Danish Testing Board) has responsibility for monitoring and evaluating laboratories for their compliance with the Danish Accreditation Scheme requirements and the OECD GLP principles. In the area of pharmaceuticals, the Ministry of Health and Pharmaceuticals has implemented a GLP compliance program whose inspectors are focusing on research data utilized to support pharmaceutical registrations. The GLP regulations for pharmaceuticals and veterinary medicine were incorporated into Danish law on February 17, 1989.

Finland

Basically, Finland does not have a compliance program; establishment of one is not anticipated in the near future. However, they do have legislation for the registration of pharmaceuticals and pesticides.

France

The French principles of GLPs (4) were originally derived from the OECD text and were ultimately published as regulations. The French directive in May 1983 indicated that the FDA considerations were to be phased in and would take the force of law by 1986. This occurred in April 1986.

The regulatory body responsible for the implementation of these GLP standards (Bonnes Pratiques de Laboratories) is the Ministère des Affaires Sociales et de la Solidarite Nationale, Direction de la Pharmacie et du Medicament. Inspections of nonclinical laboratory facilities is the responsibility of the "Inspection de la Pharmacie" and are performed by local inspectors of the DRASS (Direction des affaires sanitaires et sociales). There is also an inspection program for industrial and agricultural chemicals and is coordinated through the R.N.E. (Réseau National D'Essais).

Germany

The German GLPs, which originally appeared in the West German Federal Register on March 2, 1983, have not been published as a law but an OECD translation of GLPs in cooperation with Austria and Switzerland. A testing facility that wishes to be certified for GLP compliance must request an inspection by officials representing the Republic of Germany. This voluntary inspection program is performed at the district level under the auspices of a state authority and coordinated by the Ministry of Youth, Family Affairs, Women and Health (5). The responsibility for issuing GLP certificates and developing an inspection program within Germany resides within the states. Presently, there is a national law requiring compliance for industrial chemicals but not pharmaceuticals. It is assumed that future agreements, i.e., MOUs, and extensions of certification programs will be inclusive of a unified Germany.

Ireland

There is brief mention of GLPs in the 1985 Annual Report of the National Drug Advisory Board, but they are not presently applied. If and when GLPs are promulgated, they will most likely appear as the OECD guidelines issued under an EU directive.

Italy

Guidelines were originally issued in a Ministerial Decree in 1981 and took into account recommendations of the OECD on the mutual recognition of GLP principles. On June 26, 1986, the Italian Ministry of Health (6) issued a three-part mandate on the principles of GLP for testing chemical substances: Part A-Principles of Good Laboratory Practices; Part B-Procedures for Verification of the Application of Principles of Good Laboratory Practices; and Part C-Guidelines for Inspections of Good Laboratory Practices and Verification of the Studies.

Following issuance of these regulations, all companies applying for a license were required to provide the Ministry with complete documentation on the tests that they intended to perform (similar to

Japan's MAFF), as well as sufficient documentation to prove that the laboratory involved conforms to GLP principles. Part C of the regulation indicated that laboratories are subject to periodic or ad hoc inspections.

Korea

Korea published its GLPs (7) on September 1, 1986. These regulations were based on the OECD version and include provisions for licensing of testing facilities.

The Netherlands

The Dutch Ministry of Health and Environmental Protection has published GLP guidelines, modeled after the OECD version, which were published in 1986 and became effective January 1, 1987. The Dutch government has also concluded bilateral discussions with both the United States and Japan in late 1988. The U.S. and Dutch bilateral agreement became effective December 20, 1988.

Norway

There are no nonclinical testing laboratories in Norway that would encourage the establishment of a GLP compliance program. Therefore, as with Finland, none is anticipated in the near future.

Sweden

The regulatory body of Sweden, the National Board of Health and Welfare (NBHW), published its GLP regulations in December 1985 (8). Like all of their European counterparts, these followed the OECD versions. Internally, the text refers to the recommendations as a commentary, but in fact, they have the full force of legal regulations.

Switzerland

The Swiss Federal Office for Foreign Economic Affairs has adopted Good Laboratory Practice Procedures and Principles based on the following information:

1. Current federal and cantonal rules and regulations of the Swiss government.
2. The OECD Principles of Good Laboratory Practices and the decision concerning Mutual Acceptance of Data in the Assessment of Chemicals adopted May 12, 1981.
3. The "Guidelines" of the Intercantonal Office for the Control of Medicaments (IOCM) for GLPs concerning nonclinical laboratory studies that are in compliance with U.S. FDA GLPs.
4. The common translation by the Federal Republic of Germany, Austria, and Switzerland regarding OECD GLP principles in the German test.

Spain

The initial references to GLP were published in the official Bulletin, November 27, 1985, as a draft. Working parties from various ministries and the industry sector labored through 1987 to establish a draft similar to the OECD version. In Spain, GLP requirements were expected to be published by July 1, 1993 under ministerial order. Until recently, the GLP principles had not been published and therefore not implemented by the Spanish government because the three ministries involved (i.e., Health and Consumption; Industry and Energy; and Agriculture, Fisheries and Food) have not decided how the responsibilities for GLP implementation and monitoring would be initiated. This was resolved and the GLP Standards were issued by Royal Decree No. 822/1993 on May 25, 1993 (BOE No. 128).

United Kingdom

The United Kingdom has made the most progress in implementing GLP regulations of the countries in the European Union.

In 1982, the Health and Safety Executive (H&SE) established an inspectorate that was responsible for monitoring the testing of industrial chemicals and assuring compliance with the GLP principles for facilities generating toxicity data destined for regulatory submission.

In April 1983, the responsibility for monitoring laboratories for compliance to GLP principles was reassigned from the H&SE to the Department of Health and Social Security (DHSS). Under the auspices of the DHSS, a GLP Monitoring Unit was established and the types of facilities to be inspected included health and environmental safety testing of agrochemical, cosmetics, food additives, and pharmaceuticals were determined. Laboratories involved in testing new industrial chemicals were also reassigned to the DHSS in May 1986 (9), thus consolidating all GLP compliance monitoring activities under one directorship.

The regulations published by the British government were an annex to the United Kingdom Compliance Programme. Essentially, the document issued November 26, 1982, detailed the inspection requirements of the program and outlined the GLP principles as developed by OECD. The annex extends the provisions of the Approved Code of Practice for the Notification of New Substance Regulations 1982 (10) to the Principles of Good Laboratory Practices as published by the Health and Safety Executive. In mid-1988, the DDHS was renamed the Department of Health (DOH) with its responsibilities essentially remaining the same.

Over the past 5 years, the Department of Health, in conjunction with industry input, has published four GLP Advisory Leaflets. The Leaflets provide additional guidance to the regulated community and, therefore, Agency interpretation for specific elements or issues to GLP. The advisories can be obtained by writing to: GLP Monitoring Authority, Department of Health, Skipton House, 80 London Road, Elephant & Castle, London SE1 6LW, United Kingdom and are identified as follows:

1. Application of GLP Principles to Computer Systems - 1989
2. Application of GLP Principles to Field Studies - 1990
3. Good Laboratory Practice and the Role of Quality Assurance - 1991
4. Good Laboratory Practice and the Role of the Study Director - 1992

Japan

To date, no other national authority, including the United States, has promulgated more enforceable regulations than the government of Japan. For this reason, it is important to review these GLP principles in some detail.

Anyone wishing to register products in Japan should be aware of some unique features of Japan's regulations. For example, a Japanese facility must have confirmation from a government agency that the study was conducted in accordance with the GLP principles. Studies conducted outside Japan also must be certified for conformity with GLP principles. Failure to obtain these certifications could mean that a product registration would not be granted (11). Thus, it is extremely important that a bilateral agreement be reached between Japan and countries desiring to market in Japan. This is needed in order to standardize the acceptance of safety data that transcends the international pharmaceutical, agrochemical, and industrial chemical communities.

In general, the Japanese GLPs follow the format of the U.S. FDA regulations as well as the OECD principles; however, they differ in their philosophical approach and the intent of their application. For example, the Japanese inspectorate concentrates heavily on two specific areas: the direct role of management and the testing facility (i.e., construction, environmental conditions, etc.). The other disciplines of the GLP are of interest, but the audit process appears to center on these two areas.

In the Japanese GLPs, the term "management" means the individual in control of the overall operation, and that individual is required to prepare protocols, master schedules, and standard operating procedures (SOPs), and to assure that deviations from these are reported to quality assurance (QA).

The facility management is the supreme authority, but delegation of these types of responsibilities is permitted as long as this is clearly shown on an organizational/function chart. Under the Japanese GLPs, the Japanese agencies focus their inspections on issues such as overall facility conditions. facility designs, environmental controls, waste disposal and associated areas supporting the test facility.

Table 2a   Japanese Good Laboratory Practice Standards

	Drugs	Agricultural Chemicals	Industrial Chemicals	
Administering Authority	Ministry of Health and Welfare (MOHW)	Ministry of Agriculture, Forestry and Fisheries (MAFF)	Ministry of International Trade and Industry (MITI) supported by the Environmental Agency and MOHW	Ministry of Labor (MOL)
Applicable Legislation	Pharmaceutical Affairs Law 145 (1960)(and as subsequently amended) and Notification No. 313 March 31, 1982, issued by the Pharmaceutical Affairs Bureau	Agricultural chemicals Regulation Law 82 (1948) (and subsequent amendments) and Notification No. 3850 August 10, 1984, issued by the Agricultural Production Bureau	Chemical Substance Control 117 (1973) (and subsequent amendments) and Notification No.85 March 31, 1984, issued by the Basic Industries Bureau	Industrial Safety and Health Law 57 (1972) (and subsequent amendments) and Notification No. 261 May 18, 1985, issued by the Labor Standards Bureau
Date GLP Published	March 31, 1982	August 10, 1984	March 31, 1984	May 18, 1985
Date GLP Effective	April 1, 1983	October 1, 1984	October 1, 1985	October 1, 1985

Definition/Scope	Non-clinical safety studies on drugs, quasi-drugs, medical devices, diagnostic devices, cosmetics, and food additives that employ animals, plants, microorganisms or subparts thereof as test systems	All toxicology studies on agricultural chemicals	Tests relating to new industrial chemicals imported or manufactured in quantities greater that 0.1 ton per company per year and not on the 1974 list of Existing Chemical Substances	Ref. Law 117 including inter-mediates used in the production of drugs and agricultural chemicals
Study Types Covered by GLP	In vivo and in vitro toxicity	In vivo and in vitro toxicity	Physical, chemical, in vivo, and in vitro toxicity studies	Physical, chemical, in vivo, and in vitro toxicity studies
Applicability	All studies initiated on or after April 1, 1983. For studies initiated October 1, 1982, and after, it applies only to parts of studies extending past April 1, 1983.	All studies initiated October 1, 1984, and after where resulting data are submitted with applications for registration after April 1, 1985.	All studies initiated after October 1, 1985.	All studies initiated after October 1, 1985.

Table 2a (continued)   **Japanese Good Laboratory Practice Standards**

	Drugs	Agricultural Chemicals	Industrial Chemicals	
Acceptance of Other GLPs	Yes, where "equal to" or more detailed and comprehensive than MOHW GLPs.	No	In principle, OECD GLPs accepted. Also require statement by relevant foreign government authority certifying compliance.	Yes
Testing Facility Registration	Not required	Testing facility petition for compliance required	Testing facility petition for compliance required	Not required
Testing Facility Inspection	Required by law when necessary or periodically	Required by law following petition for confirmation	Required by law	Not required by law
Frequency of Facility Inspections	"When necessary or periodically." Approximately every 3 years	Every 3 years	Every 3 years	Not applicable
Formal Inspection	Yes	Yes	Yes	Not applicable

Check List	Notification No. 60, November 1, 1986, issued by the Pharmaceutical Affairs Bureau	Not available to public	Notification No. 226, November 20, 1986, issued by MOHW	Not applicable
Reimbursement by Sponsor of Facility Inspection Costs	Yes	No	No	Not applicable
Facility Disqualification Procedure	Not specified	Yes	Not specified	Not specified
Facility Reinstatement Procedure	Not specified	Yes	Not specified	Not specified
Notification of Compliance Requirement	Sponsor must notify contract laboratory	Sponsor must notify contract laboratory	Testing facility must ask sponsor	Not specified
Statement of GLP Compliance	Signed by testing facility manager or study director plus government authority (foreign studies)	Signed by manager of facility	Signed by QAU personnel	Signed by QAU personnel
Master Schedule	Required	Required	Required	Required

Table 2a (continued)

Japanese Good Laboratory Practice Standards

	Drugs	Agricultural Chemicals	Industrial Chemicals	
Frequency of Study QAU Inspections	Inspect each phase of a study periodically	Inspect each phase of a study periodically	Regularly or at a time necessary according to the nature of the study	Periodically or on such occasion as required by the character of the test
Qualification of Study Personnel	Appropriate education and training or job experience	Appropriate education and training or job experience	Appropriate education and training	Study director must be a natural science graduate and have one year or more of pertinent experience
Protocol Approval/ Signature Prior to Commencement of Study Management.	Approval by testing facility management and sponsor. Signature by study director.	Preparation of protocol by testing facility management. Revisions must be approved by testing facility management. Contract labs must consult with sponsor.	Preparation by study director who must obtain testing facility management and sponsors approval. Signature by study director.	Preparation by study director. Signature by study director. Revisions by testing facility management.
Final Report Signature	By study director and all scientists engaged on study	By study director and all scientists engaged on study	By study director	By study director

Analysis of Test and Control Articles for: (1) Identity, strength, purity & composition (2) Stability	Determined by testing facility. In principle before initiation of study.	Determined by testing facility or sponsor. In principle before initiation of study.	(1) Determined by testing facility (2) Yes, under storage and test conditions. In principle before initiation of study.	(1) Not specified (2) Yes, under test conditions of the study.
Analysis of Mixtures for: (1) Stability (2) Homogeneity (3) Composition	(1) Yes, when mixed with a carrier. Determined by testing facility. In principle before initiation of study. (2) Yes, when mixed with feed. (3) Determined by testing facility.	(1) Yes, determined by testing facility. In principle before initiation of study. (2) Yes, when mixed with carrier. (3) Determined by testing facility.	(1) Yes, under test conditions of the study. In principle before initiation of study. (2) Yes, when feed is employed as a carrier.	Not specified
Do Inspectors Copy QAU Documents During Inspections?	Yes	Yes	Yes	Yes

131

Table 2a (continued)

Japanese Good Laboratory Practice Standards

	Drugs	Agricultural Chemicals	Industrial Chemicals	
Test and Control Article Retention of Reserve Samples	For studies of more than 4 weeks duration, testing facility retains reserve samples from each lot of test and control articles for 5 years after the approval of registration or as long as the quality allows evaluation.	For studies of more than 4 weeks duration, testing facility retains reserve samples from each lot of test and control articles for 5 years after the approval of registration or as long as the quality allows evaluation.	For studies of more than 4 weeks duration, retain samples from each lot of test and control articles for 10 years after notification of acceptance or until no longer stable whichever is shortest.	Yes, to be kept for 10 years if in a stable condition.
Information to be Attached to Test Data Submitted to Support a Registration Application	Information on testing facility, study staff, QAU staff, and statement of compliance	Information on testing facility, study staff, QAU staff, and statement of compliance	Not specified	Not specified
Storage Location of Study Records	Not specified	At testing facility	Not specified	At testing facility
Disposition of Archives if Test Facility Goes Out of Business	Not specified	Transfer to new owner or sponsor	Transfer to sponsor	Transfer to new owner or sponsor

Retention of Records and Data	At least 5 years after the first registration application or at least 5 years after data is submitted for drug reevaluation. Wet specimens should be kept as long as they are suitable for evaluation purposes.	At least 5 years after the first registration application or submission of supplemental data and also as long as the registration remains valid. Wet specimens should be kept as long as they are suitable for evaluation purposes.	10 years after receipt of notice that test substance is classified as a new chemical substance. Wet specimens should be kept as long as they are suitable for evaluation purposes.	10 years
Testing Guidelines	Notification No 118 issued by the Pharmaceutical Affairs Bureau, February 15, 1984. For Food Additives only, reports issued by the Food Sanitation Investigation Council, July 1965, April 1972, and September 1974.	Notification No. 4200, January 28, 1985, issued by the Agricultural Production Bureau.	EPA Notification No 700. MOHW Notification No. 1039, MITI Notification No. 61-1014. Jointly issued December 5, 1987.	Notification No. 261, May 18, 1985, issued by the Labor Standards Bureau.

133

Table 2b
Recent Japanese Good Laboratory Practice Standards

	Veterinary Medicine	Food Additives
Administering Authority	Ministry of Agriculture, Forestry and Fisheries (MAFF)	Ministry of Agriculture, Forestry and Fisheries (MAFF)
Applicable Authority	Established in accordance with Pharmaceutical Affairs Laws. Issued by Animal Industry Bureau of MAFF under Notification 3912 on October 12, 1987 and Notification 1192, March 31, 1988	Issued by Animal Industry Bureau of MAFF Notification 3039, July 29, 1988
Date GLP Published	March 31, 1988	July 29, 1988
Date GLP Effective	October 1, 1988	August 1, 1989
Definition/Scope	Non-clinical safety studies for veterinary drugs	All toxicology studies applicable to food additives
Studies Covered by GLP	In vivo and in vitro toxicity studies	In vivo and in vitro toxicity studies
Applicable to Studies	Studies starting October 1, 1988	All studies starting August 1, 1989
Acceptance of Other GLPs	Yes, with relevant government certification or equivalency	Yes, with relevant government certification or equivalency

Table 2b (continued)

	Veterinary Medicine	Food Additives
Testing Facility Inspection Requirement	Yes	Yes
Testing Facility Registration	Yes	Yes
Frequency of Facility Inspection	Unknown	Unknown
Formal Inspection	Yes, GLP investigators of Animal Drug Inspection Station of Animal Industry Bureau	Yes
Master Schedule	Required	Required
Statement of GLP Compliance	Yes	Yes

At present, six separate GLP standards cover the pharmaceutical, agrochemical and industrial chemical industries as well as veterinary drugs and food additives. The first GLPs were published by the Ministry of Health and Welfare (MOHW) on March 31, 1982, and became effective on April 1, 1983. These GLPs are applicable to toxicological safety and health effects studies. The Ministry of Agriculture, Forestry and Fisheries (MAFF) GLPs apply to toxicology studies on agrochemicals. These were published on August 10, 1984 and became effective on October 1, 1984. The two agencies responsible for dealing with industrial chemicals are the Ministry of International Trade and Industry

(MITI) , which covers tests under the new chemical substance control law of 1984, and the Ministry of Labor (MOL), which is concerned with safety studies for manufacturing environments and importation of industrial chemicals. MITI published its GLPs March 31, 1984, (effective date of October 1, 1985): MOL published its GLPs May 18, 1985 (effective date of October 1985). Table 2a lists the major aspects of the Japanese GLP standards, categorized by ministry (12). Recently, two new GLP standards that deal with veterinary drugs and food additives were promulgated. These are identified separately in Table 2b. Both standards were established in accordance with the Pharmaceutical Affairs Law and it is not difficult to see why their constitution and content resemble the pharmaceutical GLPs. What is a unique relationship to these two standards is that the authority resides under MAFF. To appreciate how the GLP principles are applied in Japan today, it is best to review them on a ministry by ministry basis.

Ministry of Health and Welfare

The Ministry of Health and Welfare (MOHW), the "Koseisho" as it is known in Japan, was the first to introduce GLPs. Within MOHW, the Pharmaceutical Affairs Bureau (PAB) has overall responsibility to implement the GLP principles and monitor compliance. Three operational units enforce the compliance program: the First, Second, and Biologics and Antibiotics Divisions.

The Good Laboratory Practice Standards, which were signed as a directive by the Director of PAB under Notification No. 313 (13), are concerned with the conduct of safety testing in drugs. The Ministry first announced these GLPs as the "Standards for the Implementation of Safety Tests of Pharmaceuticals." The standards were issued on March 31, 1982 and became fully implemented by April 1, 1983. The tests subject to the GLP standards include non-clinical safety studies on drugs, quasi-drugs, cosmetics and medical devices as stated under Article 14 of The Pharmaceutical Affairs Law. Internal diagnostic drugs are also subject to these GLP standards.

In addition to defining the standard elements found in the U.S. FDA GLP Regulations (i. e., General Provisions, Personnel and Organization, etc.), this notification also requires the attachment of additional information to the test data. The type of information required by MOHW as part of the inspection process and in support of a New Drug Application (NDA) is company history; organizational charts; research staff qualifications and credentials of senior staff members; facility building plans, including photos; equipment and instrument lists; statement by management or study director attesting to GLP compliance or, in cases of data originating in a foreign country, documents from the relevant government authority attesting or certifying that the study was conducted in a testing facility complying with the GLP principles (14). The latter requirement is impossible to meet if the foreign test data are generated in a country (e.g., the United States) where the regulatory bodies (i.e., the FDA and the EPA) do not certify laboratory compliance even though the country's regulations are as stringent as Japan's. The additional information can be presented before, during, or after the inspection and must be in Japanese.

To date, this certification requirement has not been challenged, as there appears to be a proviso in the PAB Notification No. 315, "Handling Data from Studies Conducted in Foreign Countries Among Those Required in Applications for Approval to Manufacturer and/or Import Drugs," which was also issued March 31, 1982. This document lists four additional notifications (i.e., Nos. 970, 406, 700, and 852) that describe acceptable submission criteria for foreign data. Of particular importance is the provision regarding drugs which states, "If data from studies conducted in foreign countries cannot be accompanied by the documents described in Sec. 3. 5 of the notification, i.e., a government certification statement of GLP compliance, such data will be *accepted for the present* as review data for application for approval in accordance with the old policy appearing in Notification No. 970 of the PAB, dated October 1, 1976." Section 3, item 1-4 of Notification 313 (GLP Standards) describes what information is to be attached to the test data to which the standard applies. Item 5 specifically addresses foreign data and states, documents from the relevant government organization or its equivalent certifying that the study was conducted by the testing facility concerned in accordance with this standard (or foreign GLP with contents regarded as equal to or more intense than those of this Standard) ... " Essentially, MOHW is requesting certification statements with the submission of foreign data.

To complement the MOHW GLP standard, the Director of the PAB issued a manual on the performance of inspections of testing facilities, which are to conform to the GLP standards. This document, Notification No. 400 (15), was issued on June 1, 1986 with a modified detailed checklist issued in November 1986 (16).

This manual, which is analogous to the compliance manual program of U.S. FDA, describes the following:

- Aim or purpose of the program
- Types of testing facilities subject to inspection
- Qualifications of persons performing the inspection
- Performance of the inspection, rationale, and criteria for conducting the inspection
- Inspection procedure, the actual inspection process
- Report of inspection results
- Evaluation of inspection results
- Measures that should be taken on the basis of the evaluation results of the GLP evaluation committee
- Handling of testing facilities which refuse inspection
- Performance of special inspections

The MOHW grades the testing facility after the GLP evaluation committee has completed its review (17). The results can be either: Grade A, conformity to the GLP standards; Grade B, partial "non-conformity" to the GLP standards with confirmation that improvements have been made; or Grade C, complete or partial nonconformity to the GLP standards with confirmation that tests performed at this facility are unreliable.

The MOHW is the Japanese counterpart to the U.S. FDA (18), it is important therefore, to highlight major differences of the GLP principles. The MOHW requires:

- A statement of compliance
- Attachment of detailed information packages to the test data submitted in support of a new drug
- Reimbursement of MOHW for the cost of the inspection by sponsors

- Domestic inspections usually last 5 days; foreign lab inspections usually last 3 days
- Retention of records for 5 years
- Access to QAU records and ability to copy them
- A certification statement from regulatory agencies of submitted foreign test data

There is no provision for facility disqualification as there is in the FDA GLP Regulations. To date, the MOHW has implemented an aggressive inspection program in which more than 80 domestic and 7 foreign laboratories have been inspected. Recent amendments to the original MOHW GLPs were published in Japanese October 5, 1988 (Notification No. 870). The revision included the assessment of computer systems as they apply to data capture, generating, manipulation and reporting as well as additional new responsibilities of the QAU. At the same time, the frequency of facility inspections was changed from two to every three years. An official translation is expected by Spring 1989, however, to date, it has not been issued. Preliminary review indicates these changes are similar to U.S. FDA GLP amendments issued in 1987.

Ministry of Agriculture, Forestry and Fisheries (MAFF)

Although the Ministry of Agriculture, Forestry and Fisheries (MAFF) was not the first to issue GLPs, it has been one of the more active ministries in applying them. Within the MAFF, the Agricultural Chemicals Inspection Station of the Agricultural Protection Bureau (APB) is responsible for ensuring GLP compliance.

The legislative authority of MAFF for toxicology studies conducted in support of product registration appears in the Agricultural Chemicals Regulations Law No. 82 (issued 1948).

The MAFF GLP standards in support of such testing were authorized by the Director-General of APB and published August 10, 1984, as Notification No. 3850 with an effective date of October 1, 1984 (19).

In establishing its compliance program, the Ministry took a different approach from its Japanese counterparts and from other international GLP governing authorities by establishing what is known as a "Petition for Confirmation" procedure. This requires any person or organization who wishes to have confirmation by the Director-General of MAFF to submit a petition via the Director of the Agricultural Chemicals Inspection Station. The petition is a detailed document that must be in Japanese and in a format acceptable to MAFF. The contents of the petition should include:

1. Data specific to the testing facility (i.e., founder of facility, detailed organizational charts, size of testing facility, total floor area, building layout, environmental features, equipment and instrument lists, etc.)
2. Types and numbers of toxicology studies performed in the facility during the most recent 3-years period
3. Financial status of testing facility
4. Curricula vitae of researchers, including management
5. Any other data necessary to assure testing facility compliance to the GLP standards

Producing this document together is very time consuming and costly. If the document deviates from the format required by MAFF, it will not be accepted.

After the Petition for Confirmation has been formally submitted, reviewed and accepted, the Director-General of the Agricultural Protection Bureau will authorize an inspection of the testing facility. He then will schedule the time and submit the names of the inspectors. After the inspection, the Director-General will notify the petitioner of the results of the inspection and whether or not the petition has been confirmed. This process is repeated every 3 years. Although MAFF has conducted numerous inspections within Japan, it has conducted very few abroad.

If data is submitted to MAFF from several different testing facilities, each facility must be confirmed as described above.

Currently, there have been more than 60 petitions accepted with approximately 40 inspections conducted in Japan; more than 20 of these facilities have been confirmed. Although the intent of the Petition of Confirmation is to conclude with a testing facility site inspection, it is unclear at the moment how this will affect foreign laboratories as it does not appear to be enforced.

Toxicology study data submitted to MAFF in support of agricultural chemical product registration also must be accompanied by an "information package" similar to that required for the Petitions of Confirmation. Additional documents required include:

1. Documents identifying QAU personnel and their titles
2. A statement signed by testing facility management assuring that the study in question has been conducted according to the GLP standards
3. In the case of multifacility participation in a single study, presentation of evidence explaining the relationship of each facility to the study

MAFF is the Japanese counterpart to the U.S. EPA Office of Pesticide Programs, which is responsible for the Federal Insecticide, Fungicide, and Rodenticide Act (FIFRA). It is important to highlight those areas of the GLP principles that are significantly different from those of EPA. MAFF's GLPs differ from the U.S. EPA GLPs (20) in that they include:

- A facility disqualification and reinstatement provision
- A statement of GLP compliance to be signed by the facility manager (there is no option of a noncompliance statement as required by the EPA GLP Regulations)
- Protocol signature by study director is not specified. Protocols and revisions are to be prepared and approved by management
- Retention of records by the testing facility
- Permission by MAFF inspectors to review and copy QA records

As indicated previously, two new GLP standards were published since 1988 and come under the authority of MAFF, even though they were established in accordance with the Pharmaceutical Affairs Law. These GLP standards deal with veterinary drugs and food additives.

The Veterinary Drugs GLP was issued by the Director-General of the Animal Industry Bureau of MAFF on March 31, 1988, with an effective date of October 1, 1988. Since this standard was developed in accordance with Pharmaceutical Affairs Law, they have taken on the form of pharmaceutical GLPs, which differ slightly from MAFF. Facility inspections are conducted by the GLP inspectors of the Animal Drug Inspection Station of The Animal Industry Bureau within MAFF. Due to the potential disparity of GLP applications and small and large animal status, toxicity study data using small animals conducted in a testing facility conforming to the Pharmaceutical GLPs are acceptable.

The Food Additives GLP was also issued by the Director-General of the Animal Industry Bureau of MAFF on July 29, 1988, with an effective date of August 1, 1989. Similarly, this GLP standard equates to that of the Pharmaceutical and Veterinary drug GLPs.

Relative to both standards, foreign data is acceptable as long as documents from relevant government organizations certify that the study was conducted in a testing facility practicing GLP principles.

Ministry of International Trade and Industry

The Ministry of International Trade and Industry (MITI) (21) is supported by the Environmental Agency and MOHW by interagency agreements and is responsible for overseeing the safety of industrial chemicals. The Chemical Substances Control Law No. 117 has been in effect since 1973; its purpose was to confirm the reliability of test results regarding the toxicity or biodegradability of industrial chemicals. Determination of biodegradability is the final stage of testing. If a chemical is biodegradable, no other testing is required; if it is not biodegradable, toxicity and accumulation studies must be coordinated. If these results are positive, mammalian toxicity must be performed. In this latter scenario, MOHW and the Environmental Agency may assist MITI.

As a result of international GLP implementation to assure the reliability of safety test data, the Basic Industries Bureau of MITI published its GLP standards March 31, 1984, in Notification No. 85, with an effective date of October 1, 1985.

The chemical GLP standard was modeled after the OECD GLPs and later partially amended in November 18, 1988.

This GLP standard applies to toxicology screening studies including the 4-week repeated dose toxicity study using mammals, reverse mutation studies using bacteria and clastogenic study using mammalian cultural cells, in addition to degradation studies using microorganisms, accumulation studies in fish and shellfish and all toxicological, pharmacological and phamacokinetic studies.

All scientific reports for application must be conducted and reported by the testing facility in conformity to this standard. Additionally, each facility must submit a "Petition for Confirmation" relative to GLP conformance. The petition concerning the biodegradability test and accumulation test must be submitted to the Director-General of the Basic Industries Bureau under MITI, and then the toxicity tests must be submitted to the Director-General of The Environmental Health Bureau of MOHW.

MITI also has a GLP inspection program. When a GLP inspection has been completed successfully, a certificate of confirmation is issued. This document is valid for three years, after which the facility is subject to re-inspection.

MITI has basically the same policy for accepting data that was performed in a foreign lab (i.e., if the testing was performed in a foreign listed facility and conducted in accordance to the OECD GLP, the test results will be accepted if a certificate of compliance is issued by the appropriate foreign government agency). This policy was issued by the Basic Industries Bureau in Notification No. 77, May 18, 1985.

Ministry of Labor

The Ministry of Labor (MOL) also is interested in the manufacture and importation of industrial chemicals, but it concentrates primarily on mutagenicity testing as required by the Labor Standards Bureau under MOL. These GLPs, which were published May 18, 1985 as Notification No. 261, and became effective October 1, 1985, were purposefully initiated in support of Labor Standards Bureau requirements.

EUROPEAN UNION (EU)

The European Union (EU), consisting of 12 member states (Table 3), was established primarily to control tariffs and enhance trade among its members. As this involved laboratory testing by manufacturers and importers, the EU became actively involved in harmonizing the good laboratory practice guidelines and implementation programs within the

EU. The EU Commission staff actively participated in the OECD Expert Group on GLP and has been in consultation with several countries outside of the communities to seek mechanisms for the mutual acceptance of data developed under the international GLP principles.

Table 3
European Union (EU) Member Countries

Belgium	Luxembourg	Ireland
France	The Netherlands	Greece
Germany	United Kingdom	Spain
Italy	Denmark	Portugal

On December 18, 1986, the Commission of the EU published several council directories on the harmonization of laws, regulations and administrative provisions relating to the implementation of the principles of Good Laboratory Practice and the verification of studies developed under these principles.

In effect, this would call for mutual acceptance of all data and require members to submit an annual report listing all laboratories in their jurisdiction and the extent of GLP compliance. The directive also would require the conduct of drug metabolism studies under GLP principles. Initially, under these Directives, member states were required to develop legislation and be in compliance by June 30, 1988 (22), however, that date was changed to January 1, 1989 (88/320/EEC). Additional guidance was also provided for the inspection and verification process that describes the basic principles that laboratories should follow relating to the assessment of the qualifications, training, and experience of personnel; facilities; standard operating procedures (SOPs); quality assurance programs; conduct and reporting of studies; and archiving of raw data and associated reports.

The role of the EU in future bilateral or multilateral negotiations involving EU member countries and other nations is unclear. Once the directives have fulfilled by each country and a comparable compliance program has been implemented, the Commission is likely to play some approval or concurrence role in future negotiations among its members and non-EU countries.

ORGANIZATION OF ECONOMIC COOPERATION AND DEVELOPMENT (OECD)

In 1981, after much deliberation, the OECD published a broad-based version of the principles of good laboratory practices, which was modeled on those promulgated by the U.S. FDA. To avoid trade differences in the manufacture and testing of chemicals, the OECD members (Table 4) recognized a unique opportunity to harmonize test methodology and good laboratory practices on an international basis. During 1979, under the auspices of the Special Programme on the Control of Chemicals, a group of international experts developed a document that is now called the "OECD Principles of Good Laboratory Practices" (*Environment Monograph No. 45*). As pointed out previously, these OECD GLP principles have been the guiding light in transcending international boundaries both culturally and bureaucratically. This document has been, and is being, used by most countries as the model for the development of national GLP principles.

Table 4
Organization of Economic Cooperation and Development (OECD)
Member Countries

Australia	Greece	Norway
Austria	Iceland	Portugal
Belgium	Ireland	Spain
Canada	Italy	Sweden
Denmark	Japan	Switzerland
Finland	Luxembourg	Turkey
France	The Netherlands	United Kingdom
Germany	New Zealand	United States

Additionally, the expert group was also asked to develop guidance for monitoring procedures and conducting laboratory inspections for GLP regulatory authorities. The two documents that were the culmination of that effort were entitled, "Guidelines for Compliance Monitoring Procedures for GLP" and "Guidance for the Conduct of Laboratory Inspections and Study Audits." They were released under *Environment Monograph Nos. 46* and *47*, respectively.

To further enhance international understanding of critical or controversial GLP issues, the OECD orchestrated a series of successful consensus workshops. These workshops were originally intended for only government participation but it was recognized that to be fully beneficial, industry representation had to be present. This was accomplished through a selection process and at the invitation of their respective government representatives. There were three consensus workshops as noted below.

October 16-18 1990 - Bad Durheim, Germany

- GLPs and the Role of Quality Assurance
  (*Environment Monograph No. 48*)
- Compliance of Laboratory Supplies with GLP Principles
  (*Environment Monograph No. 49*)
- Interpretation and Application of GLP Principles to Field Studies
  (*Environment Monograph No. 50*)

May 21-23, 1991 - Vail, Colorado

- The Application of GLP Principles to Field Studies
  (*Environment Monograph No. 50*)

October 5-8 1992 Interlaken, Switzerland

- Application of GLP Principles in Short Term Studies
  (*Environment Monograph No. 73*)
- The Role and Responsibilities of the Study Director in GLP Studies
  (*Environment Monograph No. 74*)
- Application of GLP Principles to Computer Systems

The three consensus documents produced from the October 16-18, 1990 Workshops can be obtained by writing OECD, 2 Rue, Andre Pascal, 75775 Paris, Cedex 16 France.

U.S. EFFORTS TO HARMONIZE GOOD LABORATORY PRACTICE IMPLEMENTATION INTERNATIONALLY

Background

During 1978-1979, the U.S. Food and Drug Administration (FDA) pursued a policy of identifying competent national authorities within other countries who would be willing to enter into bilateral arrangements for the mutual recognition of Good Laboratory Practices (GLP) compliance programs. This policy was intended to reduce FDA resources that were necessary to assure that data developed in foreign laboratories was of high quality and comparable to the type of quality data developed in U.S. laboratories. At the time, the FDA was utilizing its own personnel and travel resources to conduct foreign inspections and believed that the establishment of comparable GLP compliance programs in countries submitting data to FDA could eventually reduce the level of these resource commitments. An EPA representative was invited to one of these early informal discussions with a representative of a foreign government. During these discussions, the EPA representative noted that these bilateral arrangements could also be very resource intensive and, in reporting to EPA management, recommended that EPA resources also be devoted to pursuing multilateral arrangements. There are a number of advantages to a multilateral approach.

Resources could be conserved by conducting discussions with several countries at one time. All participating countries would simultaneously hear the U.S. position, thus reducing the possibility that the U.S. might be perceived as treating some countries differently from others. The dynamics of the group effort would be likely to accelerate participation by countries that would otherwise be slow to respond. The possibility of sharing information, expertise, and resources among several countries at a considerable savings to each country could provide a considerable incentive to finalizing the necessary multilateral arrangements.

Although multilateral agreements have numerous advantages, the EPA and FDA continue to utilize the bilateral approach to international implementation.

In 1978, the United States committed itself to working within the framework of the Organization for Economic Cooperation and Development (OECD) for the eventual development of a multilateral mechanism for the mutual recognition of national GLP compliance programs. EPA and FDA jointly represented the United States in these GLP Expert Group discussions with one of the EPA representatives serving as the Chairman of the OECD GLP Expert Group.

As a result of the change in the U.S. political administration in 1981 and decreased interest of the EPA from 1981 to 1983, the U.S. State Department and the U.S. FDA became the leading U.S. participants in international discussions on GLP and GLP compliance programs. During this time, the FDA continued to seek bilateral arrangements with other countries (see the discussion on Bilateral Agreements below). The OECD Secretariat continues to have an interest in promoting international agreements among the OECD member countries as each country implements its respective GLP requirements and GLP programs.

AGREEMENTS

Multilateral Agreements

At present, there are no multilateral agreements on GLP implementation. The only international planning efforts known to be in progress are those of the OECD and the EU.

Bilateral Agreements

The U.S. Food and Drug Administration (FDA) has instituted memoranda of understanding (MOU) through a two-phase process.

A Phase 1 MOU involves signatures by the FDA Commissioner and the Commissioner's counterpart in the foreign country. This agreement calls for bilateral discussions and the active involvement of both countries' experts in the development of regulations and administrative procedures that are recognized as being comparable by both countries.

Administrative procedures for the recognition and exchange of inspection information and for the relationship between the countries are included in this initial state. (The FDA has been actively involved in training foreign inspectors, followed up by 2 to 3 week site visits in each participating country to verify the procedures. According to FDA staff, this has been a resource-intensive task. They have indicated that this training aspect could he an important task in any international effort).

A Phase 2 MOU is entered into when both parties are satisfied with the operation of their counterpart programs and mutually recognize and accept each other's data. At this point, each country has published regulations on good laboratory practices relative to safety studies, each assesses compliance of laboratories with these principles of good laboratory practice, and each has established satisfactory procedures for ensuring compliance with these principles. Thus, reciprocal recognition of the others country's good laboratory inspection programs and acceptance of test data have been achieved. The implementation of procedures/mechanisms for continued cooperative efforts by the affected countries has also been established. Thus, each party is prepared to inform the other about changes in their respective GLPs and inspection programs; share inspection results, including the level of compliance of inspected laboratories; perform reciprocal inspections if requested; and, consult on GLP issues.

At present, FDA has Phase 1 Bilateral MOUs with Canada, Germany, and Sweden. There are signed Phase 2 MOUs with Switzerland, France, The Netherlands, Italy, and the United Kingdom; a Phase 2 MOU with the Japanese (MOHW) is awaiting State Department approval.

The EPA has also signed, with FDA, a Phase 1 with Germany and a Phase 2 with the United Kingdom. Additional agreements have recently been signed with The Netherlands, Japan (MAFF), and Switzerland.

It is important to note that the Phase 2 MOU that was in effect with the United Kingdom has expired and both countries, i.e., the U.S. FDA and EPA and the UK DOH continue to conduct business as though one still exists. For MOUs that expire in 1993, the EU will not allow the signing of bilateral agreements by individual memberships. Such agreements can only be sanctioned with the EU Commission. The mutual acceptance of data agreement advocating acceptance of data between EU countries which have implemented parts 1 and 2 of OECD Decision

Recommendation (1989) provides a feasible framework providing satisfactory negotiations with the U.S. EPA and FDA can take place agreeing how to operate under OECD guidelines. The EU and U.S. delegates conducted such a formal meeting in June 1993 with the result of "progress moving forward." No definite agreement or MOU has been made with the Netherlands, Japan (MAFF) and Switzerland.

MOUs are generally signed as working agreements among the various agency heads of respective countries, whereas international agreements require a formal process involving signatures of high government officials, including approval from legislative bodies (e.g., the U.S. Senate).

SUMMATION OF AGREEMENTS

The U.S. FDA signed, or made progress toward signing, numerous agreements with agency counterparts, including:

Canada	Signed Phase I M.O.U. May 10, 1979
Japan (MOHW)	Signed Phase I M.O.U. April 15, 1983. Phase II awaiting State Department approval.
Sweden	Signed Phase I M.O.U. May 25, 1979.
Switzerland	Signed Phase I M.O.U. March 5, 1980 and Phase II April 29, 1985.
France	Signed Phase II M.O.U. March 18, 1986.
Italy	Signed Phase II M.O.U. December 19, 1988.
The Netherlands	Signed Phase II M.O.U. December 20, 1988.
Spain	Discussions in progress.
Germany (West)	Signed a Phase I M.O.U. in conjunction with the U.S. EPA December 23, 1988.

United Kingdom	A formal Phase II agreement was signed March 28, 1988, which included the U.S. EPA. Irrespective of this accomplishment, inspection and audit activities have been carried out between the two countries as though a formal agreement existed since 1979. Although the present agreement has expired, each party to the agreement continues to acknowledge the provisions of the original agreement. *Note*: This agreement has since expired (1993) and by EC Directive cannot be renewed as an individual member state agreement.

The U.S. EPA has also signed several agreements with agency counterparts including:

The Netherlands	Signed M.O.U. October 18, 1988.
Japan (MAFF)	Signed Phase I M.O.U. September 16, 1987.
Switzerland	Signed M.O.U. June 22, 1988.

Other International Bilateral Agreements:

United Kingdom and MOHW	Signed Phase I Agreement March 1984. Signed Phase II Agreement November 16, 1988.
France and Japan (MOHW)	Signed Phase II Agreement September 24, 1986.
West Germany and Japan (MOHW)	Signed Phase I Agreement September 4, 1986.
Switzerland and Japan (MOHW)	Signed Phase I Agreement late 1988.
United Kingdom and Japan (MAFF)	Negotiations for Agreement

CONCLUSION

International GLP Considerations

The United States has conducted and will continue to conduct laboratory inspections and study audits in foreign testing laboratories that develop data for submission to U.S. regulatory agencies. This policy is applicable to both FDA and EPA regulatory needs, although it may be altered as a result of current international discussions. The United States has been participating in multi-lateral discussions with several nations through such organizations as the Organization for Economic Cooperation and Development (OECD).

The EPA administrator has indicated full U.S. support for the OECD Secretariat to take the necessary steps for the development of an OECD framework to ensure recognition of GLP compliance procedures among OECD member countries. The United States believes that bilateral or multilateral arrangements among OECD national authorities should be developed for mutual recognition of GLP compliance procedures once the necessary compliance activities are in place and verifiable. As further discussions evolve, the continued development of international understandings and/or agreements will result in a U.S. commitment to developing a program to assure the quality of data being developed in certain domestic laboratories in order to satisfy the GLP compliance needs of other nations. This would also lead the United States to share its GLP compliance experiences with other nations through formal and informal communication channels. Through these international commitments, the United States will also be able to request foreign governments with comparable GLP compliance programs to assure the United States that their domestic laboratories also are in compliance with the regulatory needs of the United States. These agreements will more efficiently utilize each country's resources by minimizing costs in time and international travel. There will be a need to continue the international dialogue to assure each country involved that the comparability of their respective GLP compliance programs is being effectively maintained.

The U.S. EPA is committed to harmonizing approaches and procedures that will not compromise the quality of data developed, but will minimize testing costs and enhance the prospects of the international chemical trade. In these times, no country can consider itself immune from the realities of economic interdependence. No country should bear the burden of duplicative chemical safety testing, yet appropriate studies and procedures must be developed to help ensure that the chemicals on which we depend for our modern life style do not pose unacceptable risks either to the current or future generations. It is to be hoped that, within the international framework, a mechanism will be established wherein countries can rely with confidence on data provided by other countries. This can become a reality only when all involved countries have instituted comparable GLP procedures and GLP compliance programs and agree to utilize internationally comparable test methodologies, which are also key ingredients to the reliability of test data. Laboratories can be in total compliance with GLP, but the experimental design of the protocol may not be measuring the desired end points necessary to reach a regulatory decision. Proper managerial procedures can enhance the reliability of test data, but can never replace sound experimental design and execution.

The quality of toxicology testing has improved with the increased awareness of its critical use in providing information to assess public and environmental safety. The contributions that can be made by the industry and the testing community should not be underestimated, for these groups have a shared responsibility with government to provide the public with sound information. In addition, the testing community of each nation can enhance the quality and scope of its studies by working with industry and government in identifying and resolving managerial and scientific issues and by assisting or providing expertise to the testing communities of other nations.

REFERENCES

1. Decision of the OECD Council Concerning Mutual Acceptance of Data in the Assessment of Chemicals, Annex 2, OECD Principles of Good Laboratory Practice, adopted May 12, 1981.

2. Good Laboratory Practice in the Testing of Chemicals, Final Report of the Group of Experts on Good Laboratory Practice, No. 42353, Organization for Economic Cooperation, 2 rue André-Pascal, 75775, Paris (1982).

3. R. Herr, *Update of Foreign GLPs*. Presented at the Annual Meeting of Bioresearch Monitoring Group of the Pharmaceutical Manufacturers Association (PMA), Marriott Hotel, Washington, DC, July 1987.

4. Good Laboratory Practices, Ministry of Social Affairs and National Solidarity, Secretariat of State for Health, Instruction of May 31, 1983.

5. Bundesanzeiger, No. 42, Bekanntmachung der OECD, Grundsätze der Guten Laborpraxis (GLP), Der Bundesminister für Jugend, Familie und Gesundheit, March 2, 1983.

6. Principles of Good Laboratory Practices, Part One, No. 76, Official Gazette of the Republic of Italy, Rome, Wednesday, August 27, 1986; Department of Health, June 26, 1986.

7. Good Laboratory Practice Regulations for Nonclinical Laboratory Studies, KGLP, 4479, November 20, 1987.

8. General Recommendations of the National Board of Health and Welfare for Good Laboratory Practices for Nonclinical Laboratory Studies. NBHW-GLP, December 1985.

9. Good Laboratory Practice. The United Kingdom Compliance Programme, Department of Health and Social Security (DHSS), London (1986).

10. The Principles of Good Laboratory Practice, Annex 1. Published by Health and Safety Executive in support of Notification of New Substances Regulations (1982).

11. F. G. Snyder, *Japanese GLPs.* Presented at the American College of Toxicology, Washington, D.C., November 7, 1985.

12. D. McKay, *Japanese GLP Requirements.* J. Am. Coll. Toxicol., 6 (November 2, 1987).

13. Notification No. 313 of the Pharmaceutical Affairs Bureau. Ministry of Health and Welfare, March 31, 1982.

14. GLP Inspections in Japan. *Pharma Japan 960* (June 17, 1985).

15. Notification No. 400 in the Pharmaceutical Affairs Bureau, Ministry of Health and Welfare, June 1, 1984.

16. MOHW GLP Checklist, First Evaluation and Registration Division Pharmaceutical Affairs Bureau, Ministry of Health and Welfare (MOHW), Japan, November 1, 1986.

17. Outline of Japanese GLP Regulations and Its Procedures. Pharmaceutical Affairs Bureau, Ministry of Health and Welfare, Japan Pharmaceutical, Medical and Dental Supply Exporters' Association. *Japan Med. News* 155 (1985).

18. Part IV, Rules and Regulations, GLP Regulations. *Fed. Reg* 52 (172), 33779-33782, 21 CFR 58 (1987).

19. Agricultural Chemicals Laws and Regulation. *Soc. Agricul. Chem. Ind., Japan (II),* 59 (3850) (1985). Director-General of Agriculture Production Bureau, Ministry of Agriculture, Forestry and Fisheries, August 10, 1984.

20. Part IV EPA, Rules and Regulations. *Fed. Reg.,* 48 (230) 53946-53969, 40 CFR 16O (1983). Amended: *Fed. Reg.,* 54 (158) 34052-34074, 40 CFR 160 (1989).

21. GLP Standards Applied to Industrial Chemicals. Planning and Coordination Bureau, Environment Agency, Pharmaceutical Affairs

Bureau, Ministry of Health and Welfare, Basic Industries Bureau, Ministry of International Trade and Industry.

22. *Official Journal of the European Communities.*   No. L 15/29 17.1.87, Council Directive of December 18, 1986.

Chapter 4

IMPLEMENTING Glps IN A NON-GLP ANALYTICAL LABORATORY

Barbara N. Sutter

The Dow Chemical Company, Midland, Michigan

INTRODUCTION

Good Laboratory Practice (GLP) Standards have been in existence in the U.S. since the late seventies. The initial intent of the regulations, promulgated by the Food and Drug Administration (FDA) in 1978 and by the Environmental Protection Agency (EPA) in 1982, was to address the concerns created within the animal and non-clinical testing laboratories. Analytical labs, while not specifically addressed in the regulations, were expected to conduct characterizations of the test substance using many of the standard elements. This included sample accountability, equipment maintenance and calibration records, and documentation of reagents and standards used. The agency would audit the analytical lab to verify that there was sufficient evidence to support the analytical result.

In 1989, the EPA Federal Insecticide, Fungicide, and Rodenticide Act (FIFRA) standards were expanded to cover not only health effect testing, but all studies submitted to the agency that are intended to support research and marketing permits. This includes environmental and chemical fate studies, ecological effects, and product performance (when required by 40 CFR 158.640). As stated in 40 CFR Part 160.135(a), "All provisions of the GLP standards shall apply to physical/chemical characteristics of test control and reference substances designed to determine stability, solubility, octanol water partition coefficient, volatility and persistence of test, control and reference substance" (1). Section (b) stipulates which GLP standards do not apply to those physical and chemical characterization studies not stated in (a).

The U.S. is not the only country with GLP standards. Japan has six separate GLP standards addressing requirements for applications to manufacture drugs and devices, industrial chemicals, agricultural chemicals, veterinary medicines, and feed additives, as well as for occupational safety studies. In 1981, the 24 member countries of the Organization of Economic Cooperation and Development (OECD) published the Principles of Good Laboratory Practice which has formed the basis for GLP principles in the European Community. For the most part, all the GLPs are the same, but there are some significant differences that have an impact on analytical laboratories. A recent publication, *International GLPs* provides a comparison of all the major GLP standards (2). Other sources of information regarding the other GLP standards are listed in the recommended reading section.

This chapter will focus on the EPA/FIFRA GLPs and the application of these standards to a non-contract analytical laboratory. All section references refer to that standard (1). Any laboratory conducting work under GLPs must be aware of all appropriate regulations.

WHY DO GLPs?

The decision to do GLPs in an analytical lab should be made carefully. There are many contract analytical labs available that can conduct the analytical studies needed for product registration in accordance with all applicable regulations. What are the advantages for a corporate analytical facility to choose to become GLP compliant?

1. AVAILABILITY. A corporate analytical lab may generally be located near (or within) the corporate toxicology lab. This would facilitate a more expeditious delivery of samples for characterization, which would provide better control over the storage and handling of that material to prevent contamination or improper handling.

2. SCIENTIFIC EXPERIENCE. The corporate analytical lab personnel would have a greater understanding of the chemistry of the material being tested. While this may not be critical for many test materials, some materials have much more complex chemistries that preclude the use of the more routine approaches to conducting studies. The availability of experienced scientists to assist in these studies may make the use of a central analytical facility more practical.

3. COST. Due to the recent economic down-turn, some companies have hired contract labs to conduct the GLP studies. Other companies have chosen to minimize external costs and have subsequently established their own GLP compliant labs.

4. GLOBALIZATION. Many of the companies are now doing business around the world. Many countries have specific registration or notification requirements that must be met before any new chemical is imported. A company may choose to consolidate the required registration studies into central facilities to assure that all appropriate regulations are met.

Whatever the reason, any analytical lab that is typically doing analyses in a "non-GLP" mode can become GLP compliant. It takes a lot of dedication, training and commitment from the personnel and management. It is also not without additional costs and resources. Standard operating procedures need to be written, training carried out, and numerous audits conducted to monitor the lab's performance.

DEVELOPING A SYSTEM

Establishing GLPs in a lab that is totally in compliance with the regulations is relatively easy. That lab must have all the elements addressed in the standard and a program in place to verify the system is always working.

A greater challenge is when a lab has to incorporate GLPs into an operation where GLPs are not a requirement for much of the work. Basic research support, analytical support of production plants, and general problem solving and trouble shooting are usually the order of the day for most central analytical facilities. The decision to implement the full GLP regulations for that type of work does not make sense. An analyst trying to solve a plant problem should not have to worry with protocols, Study Directors, or in-progress audits. However, the next project that analyst performs on the same instrumentation may require full GLP compliance. The need to establish a workable system to assure the project is in compliance becomes the challenge.

A two-tier quality system is one possible solution to incorporating GLPs into a typical non-GLP lab. By incorporating some of the GLP standards as minimum requirements, a lab can be assured of a basic level of quality performance. The stricter documentation and auditing requirements would then be employed as needed to support the GLP study. Figure 1 shows this concept of a two-tier system.

Figure 1
Two-Tier Quality System

BASIC QUALITY SYSTEM ELEMENTS

Management Commitment
QAU Established
General Facility Inspection Done
Personnel Records Maintained
Basic SOPs Implemented:

REGULATED STUDIES
Protocol Written
Study Director Assigned
Quality Assurance Unit

NON-REGULATED STUDIES
Random Audits
General Lab Policies/SOPs Followed

BASIC ELEMENTS OF A QUALITY SYSTEM

Management Commitment

The success or failure of any Quality System ultimately falls to management's commitment. The staff must decide how the Quality System is structured based on the needs of the customer and applicable regulations. As stated in §160. 31, they must assure that:

- Adequate personnel, resources, facilities, and materials are available to meet the needs of all customers.
- Personnel clearly understand what is required of them.
- There is a Quality Assurance Unit (QAU) in place who understands their job function and have the responsibility and authority to notify management of violations to the lab policy.
- All policies are adhered to.
- As appropriate to the type of work, study directors are appointed and replaced as needed.

In order for management to meet these obligations they can not delegate the responsibility of writing policy and assuring compliance to the QAU. Management must understand the regulations and the needs of the customer and be involved in the setting of the policy. This does not require that management write all the SOPs but, they should be involved in the review and approval of the documents. This is an indication to the rest of the lab of their commitment to the system. They should also be involved in the general facility audits themselves and the review of other audit findings to assure themselves of the correct function of the QAU and the lab.

Management has another vital role. Too often, the announcement of the "new System" or "latest program" is delegated to a spokesperson and that individual becomes the focal point for criticism. The phrase "don't shoot the messenger" becomes important. In order for the system to become successful, management must acknowledge the "growing pains" and assist the implementation by being visible in their support.

Quality Assurance Unit

Any basic quality system must have some type of Quality Assurance Unit. Even without the GLP standards, every lab has some policies and procedures. These might include how reports are written and results relayed to the customer, how data books and raw data are documented, or records retention rules. Through the conduct of audits, the QAU can assist management in verifying the system is being followed. As part of the GLP requirements, the QAU will also have the responsibility to verify that any GLP study is being conducted in accordance to the regulations. Those duties are outlined in more detail later in this chapter.

This position in the lab can become an interesting one. As discussed earlier, management may look upon the QAU to be the spokesperson. Lab personnel may consequently focus their frustration at these individuals. The QAU must always remember to maintain a sense of humor and a positive attitude. Change is always difficult to accept. By continually showing people the positive rewards of that change, acceptance and conversion to the new system will occur.

General Facility Inspection

The general facility audit usually includes a random check of the equipment, data storage, and sample handling areas. Do lab personnel know where their SOPs are located? Does the equipment have maintenance and calibration records? Are they up to date? If there are preventative maintenance or calibration schedules established, are they being followed? How are the samples being handled? Have the training records of the analysts been updated? This audit of the facility should not differentiate between GLP and "non-GLP" areas. Unless an area of the lab or specific pieces of equipment can be isolated for only GLP use, it becomes difficult to determine if an instrument is "GLP compliant" and can be used for a study.

Personnel

As mentioned previously, management is required to provide adequate personnel for all the work to be performed. A central analytical lab may have anywhere from a few people to a few hundred people. The regulations require "each individual engaged in the conduct of or responsible for the supervision of the study shall have education, training and experience, or combination thereof, to enable that individual to perform the assigned functions"(1). The regulations also state "Each testing facility shall maintain a current summary of training and experience and job descriptions for each individual engaged in or supervising a study." This documentation should include the following:

- Curriculum Vitae (CV): The CV is a summary of the academic and professional experience of each employee. An external auditor would use the document to verify the training of the individual, so it should include a listing of publications and external presentations; attendance at technical courses, symposia, and conferences; and membership in technical organizations. The CV then becomes useful to employees as tools for tracking their own career development. It is important to note that if a CV is used, it must be updated on at least an annual basis and be consistent in its format throughout the department.

- Training Records: In order to keep the main intent of the CV focused on the technical aspect of a person's career, some departments may choose to also use training records. These records would document the employee's attendance of in-house training programs, safety training, GLP training, etc. These records should include who the trainer was, the date of the training and the signature of the employee. On-the-job training with a mentor can also be documented this way, with the mentor and employee signing the record after the employee has demonstrated a satisfactory proficiency in a specific area. These records can be developed to fit the needs of the training.

- Job Description: This is necessary to ensure employees have clear expectations as to what will be required in their function. Each job description should be specific to each person and include what type of job knowledge and organizational and leadership skills are required as well as what responsibilities will be involved.

Whatever system a lab decides to develop to document the training of its personnel, the archival of those records must also be considered. If an individual leaves the department those training records must be kept, especially if that individual was involved in the conduct of any GLP studies. External audits may occur many years after the work was completed and there must be a record of the training of the people involved. A lab could identify specific individuals who were appropriately trained to conduct GLP studies and establish the documentation necessary for compliance. These individuals would then be the only ones allowed to perform the GLP studies. However, the benefits of having established training programs, and documentation of that training can benefit all, and not just the "GLP analyst." This would also enable management to select the individuals to run the experiments based on their scientific expertise and not the presence of documentation. A note of caution: the Quality Assurance Unit individuals as well as the Study Directors assigned to particular studies MUST be specifically trained in the appropriate regulations.

IMPLEMENTATION OF BASIC STANDARD OPERATING PROCEDURES

A major part of a two-tier system is that minimum policies be established and followed for all individuals. These policies can be handled through the writing and implementation of Standard Operating Procedures (SOPs). The GLPs specify (§160.81) that certain SOPs be established. These are the ones that form the basis of a common lab policy. These SOPs include (but are not limited to):

·   Receipt, identification, storage, handling, mixing and method of sampling of the test, control, and reference substance.
·   Laboratory or other tests.
·   Data handling, storage and retrieval.
·   Maintenance and calibration of equipment.

The regulations also include a requirement for other SOPs related to the care and handling of live test systems. For an analytical lab, these are not applicable and will not be discussed in this chapter.

The SOPs mentioned above cover the fundamentals of a lab's work. At a minimum, the lab should already have established systems for receiving, storage and handling of the samples and standards brought into the lab, how work is conducted, and how the data is to be reported. These policies and procedures can be the foundation of the "minimum requirements" the lab establishes. Using the GLP standards as a guide, the lab can then enhance their non-GLP work processes to raise that minimum standard.

Each piece of vital equipment should have a maintenance record so that any analyst can verify the working condition of that instrument, and there should be policies for handling, storing and retrieving the data generated.

HELPFUL HINTS FOR WRITING SOPs

1. The first SOP that should be written is the SOP on how to write SOPs. This will provide a standard for all SOPs to follow. This SOP should address:

    ·   A Numbering System for the SOPs. A typical analytical lab may have several hundred SOPs. Consecutive numbering of all the SOPS will become cumbersome and not very helpful to the analysts. A better system would be to code the SOPs by group or function, i.e., PRO for all procedural SOPs, or QAU for QAU procedures.

    ·   A Specified Format for Each Type of SOP. The lab may have different types, such as one for policies, and another for procedures. Each format should be detailed enough that anyone can write a SOP.

    ·   A Review and Revision Process. All the SOPs need to be reviewed and revised as needed. This should be a living system, so things are constantly changing. The process for handling these changes, and who is responsible for them should be stated.

2. SOPs should be written by the people who use them. Policy SOPs should be written by those individuals, such as the management and QAU, who are familiar with the intent of the policy. Procedures should be written by the analysts who perform the tests. If they are reluctant to write them, there are a number of innovative ways to accomplish this. Have a recorder follow them around, documenting their activities, or have the analyst use a tape recorder and dictate each step as it is performed. A member of the clerical staff can then type the SOP into a final form.

3. The SOPs should be available. This can be accomplished by putting the SOPs on a computer system or having hard copies available. There are advantages and disadvantages to both approaches. The computer may not be accessible to anyone, nor is it very portable or practical in all lab settings, but the analyst can always be assured of having the most current version available. Many labs employ a system of having both paper and computer copies available, but it is important to have a way to determine the most current version.

4. Write SOPs to be used as a training document. While the regulations require the SOPs to be written, it is up to the lab to decide how much detail to put into them. If it is intended to be used by an inexperienced analyst, it should be very detailed. If the lab consists of highly qualified professionals following nationally recognized methods or procedures, have the SOPs reference those methods. In some labs there may be several different types of SOPs to handle the different uses. The emphasis is to develop a procedure that is useful to the analysts and not just as a requirement to satisfy an external agency or customer.

Frequently, people are hired into a lab, sent through rigorous safety training and then handed projects to begin working on, with little or no guidance on how to approach a project. Just as in safety, the quality of the work is important. Some labs will assign mentors to the incoming personnel, but that mentor may have a different opinion on what a data book should contain versus the mentor working with another new employee. Having a list of SOPs that each employee must read and follow, as well as a specific training plan, will assure the individual is getting the same message as everyone else. The documentation of that training is just as important as the actual training. This provides a mechanism for the training to be updated to reflect changes to technologies or policies. There is nothing so frustrating as being told the data book was handled incorrectly and the report is incomplete and is formatted incorrectly. Getting a degree from "the school of hard knocks" is extremely unproductive for the individual and the lab.

GLP REQUIREMENTS FOR PHYSICAL AND CHEMICAL CHARACTERIZATION STUDIES

As discussed in the introduction to the chapter, "All provisions of the GLP standards shall apply to physical/chemical characteristics of test control and reference substances designed to determine stability, solubility, octanol water partition coefficient, volatility and persistence of test, control and reference substance." However, this does not include all types of physical and chemical characterization studies. The regulation continues by stipulating what parts of the standards do not apply to the other physical and chemical characterization studies. This distinction among the types of physical and chemical characterization studies is unique to the EPA/FIFRA regulations.

It is easy to see why it would be difficult to understand and remember when a specific GLP standard does, or does not, apply. For example, if an analyst was conducting an octanol water partition coefficient study one month and a melting point determination the following month, both studies would require a protocol and Study Director, a QAU would need to be in place and both would require some type of inspection. However, the melting point study would not have to be listed on the Master Schedule, the protocol and final report would not have to meet all the requirements and the QAU is not required to keep copies of the protocol or master schedule. Also, there is a greater flexibility in what or how the QAU is required to inspect the melting point study. If a lab is conducting both types of studies and chooses to follow the regulations exactly, then the QAU must be very careful in keeping the documentation. However, if again, one looks at what the purpose is instead of what is required, it is easier to establish a system. Depending on the number and type of studies a lab performs, it may be more practical to adopt the complete GLP regulations for all physical and chemical characterization studies rather than trying to keep them segregated. A decision should also consider which regulations the lab must comply with, since other countries and agencies have different requirements.

Protocols

The FIFRA GLP regulations stipulate that each study must have an approved protocol and specifies, at a minimum, what it should contain. Those requirements differ for the studies depending on if they are designed to determine stability, solubility, octanol water partition coefficient, volatility and persistence (referred to here as "GLP" studies) or other physical chemical characterization studies (referred as "GLP-lite" studies). Table I on the following page outlines the differences as they relate to the protocol.

As can be seen in the table, only three excluded requirements: (8) the experimental design description, (12) the type and frequency of tests, analyses, and measurements to be made, and (15) a statement of the proposed statistical method to be used have any applicability to a physical and chemical characterization study. It is a simple matter to create a form listing all required and applicable GLP requirements as a tool for writing and reviewing protocols. Therefore, for those "GLP-lite" studies, the non-applicable items can be so indicated. In some studies the type and frequency of the analyses and measurements, while not required in the protocol, assist the analysts by providing them a more detailed contract of what the sponsor and Study Director have agreed.

Table 1

Protocol Requirements
§ 160.120

	Paraphrased Section	GLP	GLP-Lite
(1)	Descriptive title and purpose statement	X	X
(2)	Identification of TC&RS	X	X
(3)	Name/address of sponsor and testing facility	X	X
(4)	Proposed experimental start/stop dates	X	X
(5)	Justification for test system selection	X	Ø
(6)	Information about the test system	X	Ø
(7)	Procedure for identification of test system	X	Ø
(8)	Experimental design description	X	Ø
(9)	Description and identification of the diet used as well as materials used to suspend TC&RS	X	Ø
(10)	Route of administration and reason for selection	X	Ø
(11)	Each dosage level of TC&RS to be administered, including frequency and route	X	Ø
(12)	Type and frequency of test, analyses & measurements to be made	X	Ø
(13)	Records to be maintained	X	X
(14)	Date of approval of the protocol by sponsor and the dated signature of the study director	X	X
(15)	Statement as to the proposed statistical method to be used	X	Ø

TC&RS   Test, Control, and Reference Substance
X   Required
Ø   Not Required

Study Director

A Study Director is defined as the "individual responsible for the overall conduct of the study" in the EPA /FIFRA and TSCA GLPs and OECD/ GLP Principles. The Study Director's responsibilities are the same for both types of physical and chemical characterization studies and can be found in § 160.33.

Quality Assurance Unit

The FIFRA GLP regulations require that each testing lab establish a Quality Assurance Unit (QAU). This QAU is responsible for "monitoring each study to assure management that the facilities, equipment, personnel, methods, practices, records and controls are in conformance with the regulations..." The GLPs also dictate that the QAU be independent of the work being audited or the personnel conducting that study. This does not mean the QAU must be a full-time position. This would be especially difficult and costly for a small lab. What it does require is that any person acting as a QAU for a particular study is independent of that work or the personnel involved. They may be an investigator on another study that will be audited by another qualified person. The management will have to be very clear in describing how this type of QAU will operate, who they report to when acting as the QAU, and what responsibilities they have in this capacity.

The GLP requirements for the QAU are lightened for the "GLP-lite" studies as shown in Table II. The QAU does not have to maintain a copy of the master schedule or protocols for these studies. Nor does the QAU have to conduct in-progress audits of these studies, maintain records, or submit written status of findings to management and the study director. However, § 160.35 (a) does require that the QAU must "conduct inspections and maintain records appropriate to the study." This does provide the QAU with some latitude, but some type of inspection is required.

Table 2

Quality Assurance Responsibilities
§ 160.35

	Paraphrased Section	GLP	GLP-Lite
(a)	A testing facility shall have a QAU...The QAU shall conduct inspections and maintain records appropriate to the study	X	X
(b1)	QAU shall maintain a copy of master schedule...	X	Ø
(b2)	QAU shall maintain copies of all protocols	X	Ø
(b3)	QAU shall inspect each study...and maintain written and properly signed records...	X	Ø
(b4)	QAU shall periodically submit to management and study director written status on each study...	X	Ø
(b5)	QAU shall determine that no deviations from approved protocols...were made....	X	Ø
(b6)	QAU shall review the final report...	X	Ø
(b7)	QAU shall prepare and sign a statement...	X	Ø
(c)	Responsibilities and procedures applicable to the QAU...	X	Ø
(d)	An authorized employee... of EPA or FDA shall have access to the written procedures...	X	X

X   Required
Ø   Not Required

The purpose of the QA review, whether as an in-progress audit or the final report review, is to verify that the data represented is correct and defensible. Even if these reviews were not required for a melting point determination study being sent to the agency, wouldn't it make sense to do it anyway? If forms, checklists and templates are in place for one type of study, they can easily apply to all studies. If the lab is conducting studies that require the QAU to write their own SOPs, why not use them to assist in all the studies? These QAU SOPs should include:

- · Responsibilities of the QAU.
- · Protocol procedures.
- · Conducting in-progress and final report audits.
- · Conducting general facility inspections.
- · Master Schedule procedures.
- · QAU records handling and storage.

The QAU personnel need to be trained in all the applicable regulations to know when the regulations are applicable, and to guide management in appropriate interpretations. This training is available through a number of consultants in the GLP arena as well as through national organizations, such as the Society of Quality Assurance (SQA). The SQA also holds national and regional meetings that are useful for members of a QAU. These meetings provide an opportunity to hear how other companies approach the quality assurance issues and to discuss the regulations with members of the agencies.

NON-GLP REQUIREMENTS

Random Audits

Once a Basic Quality System is in place for the entire lab it will need to be monitored. Non-GLP studies do not require audits, either in-progress or final report, but without them, there is no monitoring mechanism. These audits can be very informative for all involved.

A very easy way of assessing someone's performance is to try to reconstruct their work. Randomly select a sample or final report and reconstruct the events. Was the sample tracked through the proper system? Are the methods, equipment and operating conditions stated or referenced? If another document was referenced, was it followed? Are calculations understood or given? Can the results be substantiated?

These reviews should always be conducted in a non-threatening mode. The purpose is not to catch someone doing something wrong, but to determine if they understand the current policies and are working under them. The results of these assessments are useful in understanding what additional training is required or what policies need to be reclarified. If people do not follow a SOP, find out why. Is it too restrictive? Is it impossible to follow? Is it being ignored? The audit team should then take the findings back to management. The decision of what to do to correct the situation will need to come from management and those who wrote the policy or procedure.

The team for these non-GLP assessments would not have to be a member of the QAU, but it is helpful if at least one individual in the team has a detailed knowledge of the policies and procedures. The other members could be other employees of the lab. The assessment process, if handled correctly, is a valuable learning tool for all participants. It provides a forum for discussing the policies and learning how others handle them. There may be several innovative ways to handle the same problem and an auditor may gain some useful insight on how to improve their own work.

Hint

The audit process in this case would need to be carefully standardized, especially if different people rotate through this assignment. The use of forms becomes extremely helpful and necessary to be able to compare results throughout the lab.

Benefits

Throughout this chapter many of the benefits of a quality system have been discussed. Some are tangible and measurable, such as less rework, greater productivity, reduce downtime of instruments, and faster sample turn around time. Others are less tangible. Many labs see greater customer satisfaction and happier employees as a quality system is implemented. Customers have greater confidence in the quality of the work from the lab. Employees with clearer job expectations and specific processes to follow find more time can be spent in conducting analyses and not trying to understand the latest process. They also have greater confidence in their results.

By establishing a minimum policy, the number of repeat analyses, due to lack of documentation or instrumentation problems, will decrease. It will also reduce the number of hours "reinventing the wheel" due to the lack of documentation for solving problem similar to one previously resolved.

Risks

Implementing GLPs in a traditionally non-GLP lab is not without certain risks. Both the EPA /TSCA and EPA /FIFRA regulations address the effect of non-compliance (§ 792.17 and § 160.17, respectively). If a lab submits a study that is found to be in non-compliance, it can lead to a rejection of the study, suspension or cancellation of the permit, and possible criminal and/or civil penalties (3). If the test substance used in a longer study (e.g., 2 year animal dietary study) is characterized and later found to be erroneous, the consequences can be costly. Long-term studies may be invalidated and the sponsor will have to have the study redone.

IMPLEMENTING THE ISO 9000 STANDARDS

The ISO 9000 standards are another type of Quality System. They include many of the same elements that are addressed in the GLP standards. Requirements for equipment maintenance programs, calibration records, training procedures and documentation, management commitment and the establishment of a QAU (called management representative in the ISO standards) are only a few of the common elements (4). Perhaps the major difference between the 2 types of systems is the GLP standards focus on documentation to verify the data generated, while the ISO standards focus on the system processes. While, compliance to the GLP standards is not a requirement to receive ISO certification, any lab that is following the GLPs should not require a great deal of effort or preparation to obtain certification. They will most likely discover that the ISO standards enhance an already strong Quality System.

CONCLUSION

Any number of systems can be established to address the varying "levels of GLP standards" required for each type of physical and chemical characterization study. It is important to pay attention, however, to the purpose of conducting the studies under GLP and not look for "the easy way out." To gain the greatest benefit from any Quality System, the analytical lab must implement a system that is easy to follow, clearly understood by all, and one that is designed to lead to the success and productivity of all involved. A system that has been designed to work for the lab, and not the other way around, will have a higher degree of success.

REFERENCES

1. Environmental Protection Agency, *Federal Insecticide, Fungicide and Rodenticide Act (FIFRA); Good Laboratory Practice Standards* 40 CFR 160. Final Rule appeared in the *Federal Register* 54:34052-34074, August 17, 1989.

2. DeWoskin, R.S. and Taulbee, S.M. *International GLPs*. Interpharm Press, Buffalo Grove, Illinois, 1993.

3. U.S. Environmental Protection Agency, Enforcement Response Policy for the Federal Insecticide, Fungicide and Rodenticide Act Good Laboratory Practice (GLP) Regulations; Office of Compliance Monitoring; Office of Pesticides and Toxic Substances; September 30, 1991.

4. Kracht, W. R. "Implementing the ISO 9000 Quality Standards in a GMP or GLP Environment", *Implementing International Good Practices*; Dent, N. J., ed.; Interpharm Press, Buffalo Grove, Illinois, 1993.

RECOMMENDED READING

1. *Implementing International Good Practices*; Dent, N. J., ed.; Interpharm Press, Buffalo Grove, Illinois, 1993.

2. Garfield, F. M.; *Quality Assurance Principles for Analytical Laboratories;* 2 Ed; Association of Official Analytical Chemists, 1991.

3. Hirsch, A.E., ed.; *Good Laboratory Practice Regulations*; Marcel Dekker, Inc., New York, 1989.

4. Carson, P. A. and Dent, N. J., ed.; *Good Laboratory and Clinical Practices,* Heinemann Newnes, Oxford, 1990.

Chapter 5

GLPs IN AN AUTOMATED LABORATORY

John J. Fitzgerald, Ph.D. and James N. Bower, Ph.D.

Automated Compliance Systems, Inc., Bridgewater, New Jersey

INTRODUCTION

There are several dangerous words used in the discussion of proper and optimum lab operation. Good Laboratory Practices (GLPs) are required for labs performing non-clinical studies under 21 CFR 58. The implication of the word "good" is that all labs would be better, closer to the ideal, if these precepts were followed. The connotation is extended with the U.S. EPA's Good Automated Laboratory Practices (GALPs) and the FDA's Good Manufacturing Practices (GMPs). A cultural assumption in Western civilization since the introduction of the machine age, one that has been elevated almost to a religious tenet since the 1960's, is that "automation" is better than the alternative non-automation.

The lab director is challenged to merge the requirements of what is mandated - GLPs, GALPs, GMPs - with what is thought desirable -- automation. In the real world laboratory, opportunities are seldom absolute. There are often resource, budget and technology constraints which require each organization to develop flexible plans to address their operational goals. In those labs which are regulated by the "G's," economic criteria are not alone sufficient to make design decisions. On the other hand, few regulated labs can implement the "G" regulatory precepts without some regard to financial issues. Reasons for automation can include both procedural confirmation of good practices and financial benefit. Your lab can achieve procedural confirmation without automation.

This discussion introduces a measure, economic behavior, to evaluate and optimize the use of automated and instrumented technology in regulated labs. The approach describes a generic lab model with detailed information flow. Each informing process is evaluated against the GLPs and automation opportunities, and an index (called the GLP Economic Behavior Index) calculated to assist the lab designer/director in prioritizing implementation.

Before economic behavior can be addressed, we introduce a model of lab internal operations which is built on the concept of analytical operations (ANOPs). ANOPs are then separated into their sub-components: chemical operations, measurement operations, and data operations. Each may be the subject of automation, and all are subject to the GLPs. Automation of chemical and measurement operations has progressed quite far at this time. Therefore, we use data operations automation as the subject of a discussion of requirements for GLP monitoring, auditing, and process confirmation actions.

The automation model presented here has implications for system confirmation and validation. These implications are explored separately in order to understand the limits of confirmation of automated informing processes.

PROCESS CONFIRMATION GOALS FOR AUTOMATION

If we take the financial goals of automation as self-evident, a focus on the GLPs leads inevitably to the requirement for process confirmation. The confirmation of lab processes is required by the CFR references to both ongoing process monitoring and the regulatory audit. If we take a critical look at the regulations, a list of desirable effects of automation projects is immediately evident:

- **Isolation/Reduction of Human Intervention** - The ultimate goal for lab automation would be to remove any need for human intervention or interpretation from the processing of data. At the very least, an automation system should reduce the scope of human intervention to decisions (e.g., pass/fail) or the capture of data which cannot be captured automatically. Once a system has been validated, this would allow a tighter focus on the inevitable problems which appear in the lab when humans make mistakes, and on engineering designed to reduce such mistakes.

- **Strengthened Automation System Specifications** - Traditionally, data automation systems have not been very capable. Labs have been forced to deal with multi-step, software-assisted manual data reduction processes. Part of the reason for this has been the inability for automation systems to communicate with other systems. Another part has been the inability to properly specify automation tasks to the level of detail necessary to complete data processing in even the simplest of scenarios. A strengthened specification process would allow more capable and automated data reduction processes.

- **Protection from Deliberate Tinkering** - People respond to pressure in a number of ways. When they perceive advantages to themselves they occasionally find ways around even complex control systems. Automation systems should make this more difficult by identifying the sources of data and changes within the system.

- **Controlled, Self-Documenting Systems** - Software systems which have been validated (or confirmed) become valuable assets. They can be assumed to be working if no tampering has taken place. The control of software qualification and maintenance should be easier in a well-constructed automation system.

- **Paperwork Aids** - As surprising as it may seem, the GLPs insist on paper as a means of confirming that an automated process works. The automation system should therefore be able to contribute to the paper collection which the quality assurance unit maintains.

·   **Simplified Confirmation -** The process of confirming proper lab
    process execution should be simpler and demonstrable at the end
    of an automation project.

·   **Mechanized Audit Trails -** Audit trails are required by the
    GLPs. Data audit trails are an obvious requirement. However,
    since other "informing processes" are present in the lab, other
    audit trails are also necessary.

·   **Set up a Framework for System Validation -** Validation of
    computer software systems is a complex task. Without spending
    time on how that may be accomplished, it is necessary to state
    only that an automation system should be designed so that
    validation is easier with the system than without.

LAB OPERATIONS AUTOMATION MODEL

In order to develop a methodology for automation which is consistent
with GLPs (and GALPs), we need a model of lab operations which is
generic enough to handle labs of widely varying types and complexity,
yet simple enough to yield meaningful and timely results. In order to do
this we need a lexicon which is generic and relatively simple, and which
introduces concepts necessary for automation.

Lexicon

Elements of the lexicon may seem obvious. However, without the
definition of terms, there can be no effective discussion of the automation
model.

·   **Sample -** A collection of material taken from one place at one
    time or mixed under control from several sources for further
    measurement, analysis, or determination of contents. A sample
    may consist of several containers which are treated differently in
    preparation for different analyses, but must originate from one
    "material" (tissue, chemical substance, or other substance) or

constitute one well-described "entity" (animal specimen, container of a pharmaceutical preparation, etc.).

- **Sample Identification** - The entire collection of information which places the sample in the context of study. It is not unusual for lab systems to use a limited number of short-hand fields to approximate sample identification. However, the meaning of the term here is the complete collection of information to be used in selection or reporting of test results for subsequent further use in context. The collection of sample identification information captured from the sample source should uniquely identify the sample and its attached data for later use.

- **Test** - The procedure or set of procedures to be applied to sample material (or a subset) to measure, determine, or analyze the material. The consequence of performing a test is to generate information (test results) about the sample material.

- **Test Results** - The collection of information generated by the lab about the sample material. The test results may include numerical values, textual data, ratings, spectra, raw binary data, and more.

- **Analytical Operation (ANOP)** - The operations which make up the testing procedure. ANOPs are self-contained portions of a test method which, under the control of a Standard Operating Procedure (SOP), and under the guidance of trained personnel, use appropriate information, procedures, and materials to generate intermediate or final information for the test.

One or more ANOPs may be required to perform a test. The procedures can be manual or automated. Portions of an ANOP may be physical, chemical, instrumental, or data oriented.

ANOPs take sample material, sample information, information from other ANOPs, and information from the ANOPs of other samples as inputs. As output, ANOPs can produce transformed information, new information (from measurements, determinations, and analyses), decisions, and altered sample material for use in other ANOPs.

· **Chemical Operation (CHEMOP) -** An operation which chemically transforms the sample material to make it more suitable for analysis. Physical transformations are included here for the sake of simplicity. CHEMOPs are also performed under the guidance of an SOP by trained personnel. CHEMOPs may produce intermediate or final information such as weights, volumes, spiking information, etc.

CHEMOPs may be partially automated. The traditional automation of CHEMOPs involves equipment such as robots, multi-sample agitators, automated syringes, etc.

· **Measurement Operation (MEASOP) -** An operation which measures, analyzes the content of, or determines the value of a component of sample material. MEASOPs generate data which may require additional processing for its final use. MEASOPs require SOPs and trained personnel.

MEASOPs are traditionally automated by instrument manufacturers with devices like robots or auto-samplers, and with data station software for instrument data reduction. Usually these systems cannot provide final data reduction, since the data station software typically cannot be driven from the central lab system, and since tools for data reduction beyond traditional instrument automation boundaries are not usually supplied.

· **Data Operation (DATAOP) -** An operation on data alone, resulting in partial or final reduction of the data to the usable state. Manual DATAOPs require SOPs and trained personnel. When DATAOPs are automated they can be performed without human intervention, under the control of standard automation procedures (SAPs).

· **Data Group (DATAGRP)** - A collection of data which is generated or collected at one time, and which pertains to one sample, test, and measurement execution. Examples of DATAGRP are single Ph readings, a multi-compound quantitative data file from a gas chromatograph, or a series of replicate measurements collected from an automated wet chemistry analyzer.

Definition of the Model

Prior to this point we have mentioned using a GLP Economic Behavior Index[1] (GLP EBI) to identify which tasks should be automated. However, the purpose of the GLPs and the attendant quality assurance efforts is only to establish, then confirm, the appropriate operation of lab processes. In order to do this, we need to understand how analytical operation automation can work to reduce and isolate the traditional human weaknesses inherent in any process. Our model for this effort first establishes how ANOPs work to perform lab processes, then breaks down ANOPs into their components. As mentioned above, the CHEMOP (chemical operation) and MEASOP (measurement operation) have long been the subject of automation development by equipment and instrument manufacturers. The DATAOP (data operation) component, on the other hand, has resisted automation for many years. The proper tools are just becoming available to implement DATAOPs in compliance with GLP requirements. Our model will use the DATAOP as the subject for a discussion of requirements.

ANOPs in Lab Processes

Our first task is to understand how ANOPs can be used to perform the work of the lab. Our lab model allows each test to be performed by a series of ANOPs, each under the control of a Standard Operating Procedure (SOP). Each ANOP can use information from the sample,

---

[1]Automated Compliance Systems, Inc. Copyright © 1993

other ANOPs, and other samples. In addition, it has access to materials (original or altered) from the same sample. The purpose of each ANOP is to accomplish a logical subset of the overall methodology for the given test. Since tests are often specific to the sample matrix or material, individual ANOPs and the collections of ANOPS used to execute a test methodology are also specific to matrix.

A general picture of the information inputs and outputs for ANOPs is displayed in Figure 1. It is quite clear that sample materials are used within analytical operations, and that new information is created. However, it is also necessary for ANOPs to share information from within a sample and between samples for successful automation.

Figure 1 Analytical Operation Inputs and Outputs

[Altered] sample material	ANOP transformed information
Sample information	New ANOPs information
Information from other ANOPs	Decision(s)
Other sample ANOPs information	[Altered] sample material(s)

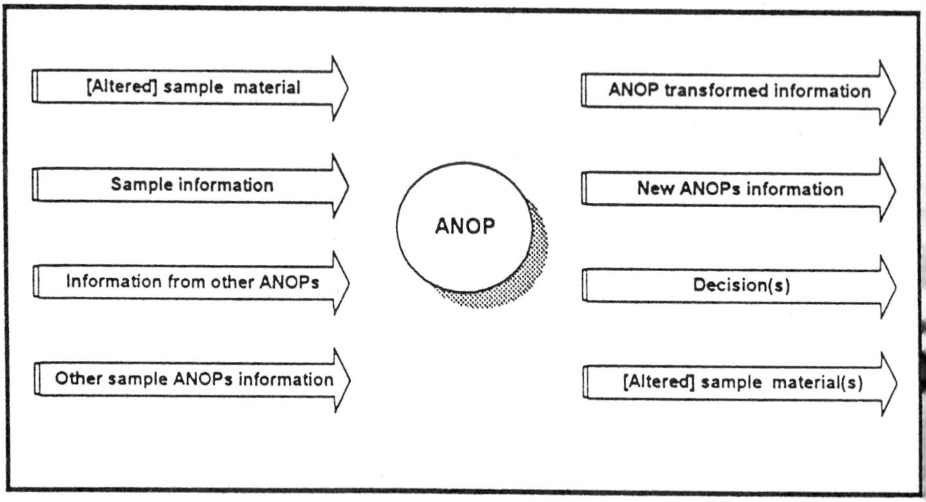

An example of ANOPs for a sample with two tests scheduled is depicted in Figure 2. These two tests share some information. An example might be the sharing of data from a "weight loss on drying" or "% moisture" test with an analytical determination of an active pharmaceutical component for the purpose of calculating a dry weight assay. In addition, some steps in our example generate new information.

An example might be a chemical preparation step which must record preparation step weights and volumes. Some steps simply transform existing information (as spike recoveries are calculated from original sample concentration, spike sample concentration, and spiked concentration).

At this level, ANOPs are nothing more than normal lab procedures, executed in various lab departments, connected normally by a paper trail which would be the subject of GLP process confirmation efforts. In order to proceed with automation, we need to break down the internal activities in an individual ANOP and take a look at requirements for automation.

Figure 2 Example of ANOPs for a Sample With Two Tests

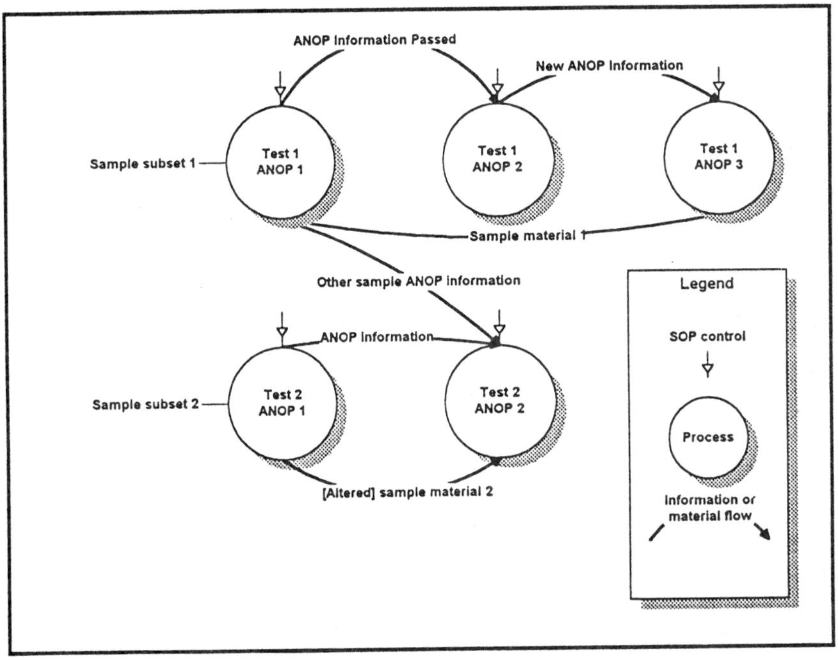

Decomposition of Lab Processes

An individual ANOP is composed of a series of CHEMOP, MEASOP, and DATAOP processes (Figure 3), each of which requires control by an

SOP. An ANOP does not necessarily require all of these operations, but must have at least one. The path through an individual ANOP can vary with the methodology. It can be quite simple, requiring only a DATAOP, or quite complex, with several operations of different types. At the most complex, an ANOP could have alternative pathways through the processes, when options may exist for alternate chemical, measurement, or data operations depending on circumstances.

Figure 3 Analytical Operation Decomposed

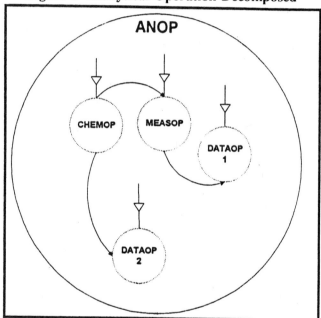

All elements of ANOPs require SOPs for guidance. At this level, SOPs must specify physical and chemical transformations, measurements, and data transformations. We will need a new type of specification to assist in the automation of DATAOPs, however.

Before automation begins we must first have a mechanism for deciding which types of automation projects are appropriate and will lead to a GLP-compliant lab.

THE ECONOMIC BEHAVIOR MODEL

The intent of the GLP regulations is to require laboratories to do their measurement and analysis business in a managed, thoughtful, planned, and proper way so that conclusions drawn from lab data are valid. Key words which show up again and again include:

- Assure
- Plan
- Maintain
- Monitor
- File
- Archive
- Describe
- Record

These are all words which deal with control of the vagaries of human organizations and scientific investigations. To date, both organizations and investigations have been managed procedurally by humans, through written documentation of and checking of procedures and audit or inspection processes. In a sense this has been necessary, since human procedures have generally not been detailed enough to guide machine systems, and since humans can adjust to changes in their environment and investigations which are too subtle, small, or unforeseen for inclusion in written or computerized procedures.

Definition of the Model

Laboratories serve as net creators or generators of information. They take in requirements, schedules, and sample material, and deliver large amounts of measured and/or transformed information about the materials. Since the labs are large generators of study data, we need to understand the implications of generating data which have impact on the GLP process.

The GLPs require confirmation of lab processes, whether this is through quality assurance monitoring or the audit process itself. Confirmation means ensuring that each process is not only properly defined, but supported adequately by management with resources, and that processes are actually performing as specified. Since process confirmation is an aspect of quality management, it is appropriate to use the language of quality to describe it. The cost of non-compliance with GLPs is well understood and enormous for the regulated community. What has not been well characterized is the cost of compliance. This cost is strongly affected by the organizational practices behind the processes. The more effective these practices are at organizing information for use in confirmation, the lower the cost of the ongoing confirmation processes which are used to determine compliance.

Confirmation processes consist of all the efforts required to gain confidence that ANOPs are proceeding according to requirements. These include the time and expense devoted to quality assurance monitoring on a regular basis as well as the costs incurred during actual regulatory audits. In addition, cost of compliance includes all organizational costs devoted to maintenance of data in an audit-ready status. For example, maintenance of tape libraries, archives, paper records, and more are all included.

As we gain experience with automation, it becomes clearer that definition of processes and procedures must become more and more detailed to address both the needs of process confirmation and automation in the lab. Better procedure definition has the effect of removing human judgement (and fallibility) from lab processes, and allowing the creation of more complete automation specifications.

For those who have been involved in implementing automated procedures and systems, it seems as if there is almost never enough detail available to successfully automate an informing process. If detail proves adequate and available, the tools often fail to be able to integrate information from multiple sources or to implement automated procedures easily. As if these problems were not enough, there is often a conflict between expressed procedural guidelines and actual practice. When these problems prevail, automation projects fail. The perceived financial benefits of automation and those of process confirmation fail to be realized.

This discussion will shed light on some simple measurements which any responsible party can make to judge the appropriateness of an automation project in view of GLP requirements. The measurements are combined into a GLP Economic Behavior Index (GLP EBI) which can be used as a measure of the suitability of lab informing processes for automation.

The Informing Processes

Since a lab is a net generator of information, it must contain at least one informing process. In addition, since it performs its business through procedures, informing processes must be present to account for the execution of those procedures. A good definition for an informing process, then, is a process which either creates, accounts for, or develops relationships between data items.

Any of the components of ANOPs can qualify as informing processes. In addition, processes like login and method requirements definition also qualify. Each of these informing processes is a target for automation. However, before beginning an automation project we need a method to help prioritize which ones to attack.

The GLP Economic Behavior Index (GLP EBI)

An obscure quote that has been attributed to Pascal notes that "all progress begins with measurement." We cannot begin to understand how to improve until we know where we are. The assessment of the suitability of procedures for automation and GLP compliance must therefore begin with measurements. "Economic behavior" is a term introduced here to describe the effects of human behavior on the economics of process confirmation. Economic behavior (or *economical* behavior) can be measured in a series of steps which result in a system for determining the appropriateness of GLP processes for automation. This measurement system is called the GLP Economic Behavior Index (GLP EBI).

Step 1: Isolating the Lab - Reducing Complexity

The first step is necessary since GLP labs can exist in a variety of scenarios. Labs can act as anything from providers of stand-alone services to labs completely integrated into the overall investigation scheme.   They can involve investigations which span many disciplines, from animal testing to chemical and physical testing. Standalone labs can even serve in multiple studies in a variety of relationships with the sponsoring organizations. They may function as a method developer for one study, an analytical services lab for another, or execute a portion of a study for a third.

For the purpose of developing the Economic Behavior model, we need to treat the lab as something of a black box with respect to its sponsors or clients. For this to succeed, the sponsors must appear as isolated sources of, and sinks for, information. That is, the lab processes must be isolated from the sources of data and from the end users of the data.   For example, study sponsors are sources of requirements definitions, sample materials, and testing plans. Sponsors are also sinks for test reports, status reports, and schedule updates. The lab is responsible for connecting the incoming and outgoing information in a proper and procedural manner. In this model, only sponsors and clients make decisions and turn information delivered by the lab into decisions and information useful in the study. In order to use the Economic Behavior model, this means that labs with complex missions such as method development must treat data generation as their product. Data use by professionals must be treated as a separate function.

Cordoning off the lab in this way helps to clarify the various informing processes present within a given study, and makes it possible to isolate and define what automation is possible.

Step 2: Inventory the Informing Processes

The natural next step is to inventory all the informing processes within the lab, with the aim of creating an information bank for understanding and applying resources to automation opportunities. Appendix A contains a very generalized informing process inventory which covers processes found within varied laboratories.

Labs may have very complex informing processes if they serve many purposes or participate in various studies in different ways. Undertaking an informing process inventory in a given lab may be more effective if it is first directed toward an individual study, then generalized to include processes which serve the remaining studies.

<u>Step 3: Define the Elements of the GLP EBI</u>

In a general sense, lowering the economic behavior cost of lab processes requires that entropy be lowered (in the chemical/disorder sense of the word), while information content is raised (in the information theory sense). The third step in building the GLP EBI involves assessing the orderliness and information content of the lab informing processes. This step requires the development of a cost basis for each informing process, on a relative scale, to support automation decisions. This means that very complex organizational behavior must be reduced to a rational model.

The first part of the definition of the GLP EBI requires understanding and breaking down each informing process into elements which can potentially add to the cost of confirmation (and the cost of compliance). These elements relate to information content or orderliness of the information process, or both.

The component elements are then assigned weights which correspond to their relative importance in the given lab organization. A relative scale for each element is assigned such that the high end corresponds to more costly confirmation requirements. Costly is defined here as more disorderly or containing higher information content. In a sense, this step corresponds to a traditional time and motion study. The costs for the GLP EBI are not related to performance of the informing process itself, rather *they are related to the performance of the confirmation of the process*.

Primary measurements for economic behavior costs are often not available, and therefore the numerical scales are somewhat subjective. Nevertheless, they are invaluable for the support of automation decisions.

GLP EBI Components – Applicability Criteria

The GLPs are constructed as a general set of guidelines which describe the activities in and results from a well-run lab. Within the application of the GLPs there is a certain amount of leeway in the mechanism of compliance. In the audit process (as opposed to confirmation processes executed by a quality assurance unit) there is a practical need for auditors to use their time most effectively and appropriately. Inside the lab there is usually an uneven effort to generate compliance from one area to the next. These can generate economic advantages or disadvantages in compliance automation efforts.

- **Certainty of Audit** - GLPs are not a betting game (unless you are somewhat demented and willing to gamble the future operation of your company). However, auditors can place more or less value on compliance in different areas. The certainty criterion is intended to provide a measure of the certainty that an audit will touch on a particular process. If you plan no confirmation activities, and anticipate none by auditors, a process probably does not require much work in preparation for audit or in process confirmation. Therefore, the process would receive a low score.

- **Audit Value** - The GLPs describe the desirable activities in labs in a series of paragraphs which deal with the goals of various personnel, activities to be performed, records to be kept, and communications to be made in the lab. Many of the lab processes which need to be confirmed are covered by multiple citations. The audit value for a process is presumed to increase if the process is mentioned or implicated several times in the CFR. Higher numbers of citations would correspond to a higher cost of confirmation.

- **Completeness of Procedure** - A procedure which is extensively documented in an SOP, well understood by personnel, and which requires no interpretation by personnel, reduces the cost of process confirmation. Likewise, a poorly documented procedure raises costs.

Data Generation Criteria

Informing processes by definition generate or manipulate data. Therefore, one portion of economic behavior relates to measurement of how this occurs in the lab.

· **Volume of Data** - Lab processes which generate high volumes of data can also produce mistakes at high volumes. There is a presumption that automated equipment produces fewer mistakes. On one level this is true. Fewer manual operations leads to fewer human mistakes. However, production of large volumes of data often leads to the use of automation systems based on data workstations. In turn, data collection automation systems require proper setup, correct sample identifications, and the creation of valid relationships between samples. Humans participate in each of these areas, causing the potential for problems.

· **Importance** - Laboratories often labor at their lowest levels without completely understanding the significance of the measurements which they make. This occurs for several reasons:

·· Dealing with large volumes of measurements leaves little time to understand the relationship of data across complete studies,
·· The details of studies are generally not known to lab workers, and
·· The blinded nature of studies protects the results from improper bias by design.

However, lab management often can assign some measure of the importance of the measurements made. The importance of lab results generated or used in an informing process should, in some measure, participate in the economic decision to automate it.

· **Rate of Generation -** In many labs, extremely significant information is generated at a very slow rate. A good example might be drug stability studies, where sampling of stored materials may last for years. Despite the importance of the generated information, confirmation would require only relatively low cost efforts.

· **Number of Transformations -** Typically, data are captured and reduced in steps, with each step devoted to a particular type of transformation (e.g., wet weight to dry weight adjustment, spike recovery calculation, and internal standard correction). When the number of data reduction steps increase, the difficulty of confirmation also increases. The increase in cost is more than arithmetic, since audit trails, transfer of information between steps, and proper routing are all subjects which must be confirmed in multi-step transformations.

· **Number of Manual Steps -** Likewise, costs increase with the potential for error introduced by human operations.

· **Transfer of Data -** Data which require physical transfer between personnel or systems for proper reduction also increase the cost of confirmation. Audit records, data transfer quality checks, and custody records all add to complexity and cost.

Information Retrieval Criteria

GLP compliance confirmation costs labs time and money. This is certainly true during an audit, but even more true during the continuing quality assurance activities devoted to confirmation on an ongoing basis. Since people costs are the basis of confirmation actions, measurements in this area deal with the amounts of time necessary to confirm the informing process.

- **Nearness** - A good deal of time may be taken in actually physically searching for information during confirmation. If it is stored near the logical starting point for confirmation, the cost will be lower.

- **Difficulty of Integration** - Informing processes are not guaranteed to be simple. In fact, as first configured, they are often cumbersome and complex. They may rely on information coming from a variety of sources, and may require judgement on the part of the person executing a task. Confirmation of a process involving integration of information from many sources and involving human judgement will necessarily be complex and time-consuming.

- **Number of Disciplines Involved** - Labs, because of their technical nature, frequently employ multiple, interrelated disciplines. When several disciplines are involved, confirmation will probably require training and expertise in those disciplines, as well as a greater amount of time.

- **Number of People Involved** - Processes frequently use people to accomplish their tasks. Reconstruction of an execution of an historical process during confirmation may require time spent with each person involved, particularly when judgements are required.

- **Frequency** - Process confirmation is presumably applied in a statistical manner (at least if it is to be defendable). Therefore, a higher informing process frequency must result in higher confirmation frequency, and therefore, cost.

- **Duration** - Informing processes take varying amounts of time (from seconds to months). Those which take a long amount of time may likewise take a longer time to confirm. Personnel may not be applied directly during the entire confirmation process. However, the longer time period may require changes of focus and context, as well as interruptions. These, in turn, cause higher confirmation costs.

· **Latency** - Process confirmation costs more when it is attempted well after the fact. Sometimes this is necessary. However, a statistical quality assurance effort can reduce the cost of confirmation by tying the confirmation process directly to the execution of the informing process.

Additional Elements

Your lab may require additional elements to record economic behavior costs which are unique to your organization and processes. When the elements have been decided, the last step in preparing for GLP EBI measurement is to transcribe the elements onto a structured form to aid in measurement. An example is found in Appendix B.

Step 4: Measure the GLP EBI for Informing Processes

The mechanics of the measurement process are quite simple. First, the names of the informing processes are transcribed onto individual copies of the GLP EBI worksheets. Then each informing process is rated on the various elements by knowledgeable personnel. Since the collection of knowledgeable personnel will vary from process to process, it may be necessary to reconcile the scoring efforts from process to process by further interviews with a scoring team.

Ratings are first totalled for each informing process, then multiplied by the weighing factor for the process to arrive at a final process economic behavior cost. The data collection and reduction efforts can be mechanized easily through the use of a spreadsheet or PC database program.

FOCUS ON DATAOP AUTOMATION REQUIREMENTS

As discussed above, DATAOPs represent the processing of data derived from another portion of an ANOP. Since CHEMOPs and MEASOPs usually act as generators of data, the initial kickoff for a DATAOP is the capture of that data. Each DATAOP is then the entire set of processes which reduce that captured data to its final format within the scope of the analytical operation to which it belongs. DATAOPs may simply transform data, or actually create new data. Some DATAOPs may involve even more complex, conditional steps which are applied only to samples of certain types (spikes, blanks, etc.), or to data of certain types (e.g., surrogate compounds, regular chemical compounds, or sample preparation data alone). Here are some examples.

- The simplest DATAOP would be the manual capture of data from a chemical or measurement operation.

- Another simple DATAOP might be the single step calculation of a weight from balance measurements: subtract the tare from the sample weight.

- A more complex example of a DATAOP might be the processing of gas chromatographic (GC) quantitative data in multiple steps:

    1. Convert microgram amounts measured by the GC data station to initial concentrations, based on sample and injection volumes.
    2. Correct concentrations by the use of internal standard recovery data.
    3. Correct concentrations to a dry weight basis using loss on drying data for the sample.
    4. Correct concentrations to a ratio using the initial pharmaceutical dosage.

Figure 4 on the following page depicts a typical DATAOP and its steps.

Before the automation of a DATAOP, these internal DATAOP processes tend to be manual in nature. They may require data transfers, use of multiple software tools on multiple software systems, and occasional data transcription between steps. It would not be unusual to see a single DATAOP use manual transcription, followed by spreadsheet manipulation, transcription into an intermediate report generator, and movement of data into a final database system.

Each of these steps requires human intervention with associated error. The complexity of such a DATAOP must be backed up with an SOP and a confirmation process which take into account the affirmation that such steps are correctly executed each and every time. Under these circumstances, validation alone would be an expensive process. Confirmation of the DATAOP process would also be very difficult and expensive.

Figure 4 DATAOP Decomposed

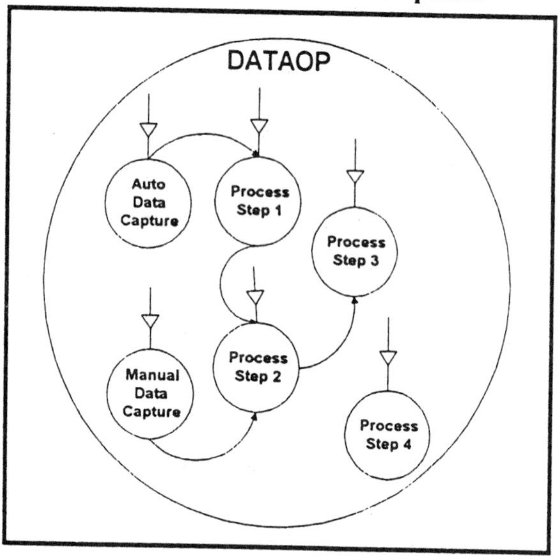

As the first stage of avoiding this expense we need to understand how the internal operation of a DATAOP can be automated. Understanding the internals of a DATAOP will make possible construction of an automation system which would have the benefits of:

· Reducing the traditional human weaknesses to tightly defined areas

· Isolating the impact of human mistakes, and

· Allowing a simpler confirmation process

Such an automation system will lower both DATAOP process and process confirmation costs.

Humanless Processing

The ideal goal in data automation is to make possible complete data reduction without human intervention. In order for hands-off automation of routine data operations to succeed, a great deal of preparation must be accomplished. For the preparations to be successful we need to thoroughly understand how to specify a DATAOP, and how to control the automation in a defendable, confirmable way.

1. **Identification of Data** - The human activities which have unalterable and potentially disastrous implications on the quality of DATAOP data processing fall into a limited set of categories. Many of these deal with the identification of samples and data or the establishment of relationships between samples (for example, quality control groups, or instrument data groups). These tasks are historically the source of many errors. The places where these errors occur are the typical man-machine boundary areas (for example, sample login, work group creation, QC assignments, and instrument and workstation setup). Simply reducing the complexity of data processing by automation will limit human intervention to these areas. This action in turn will refocus process confirmation to a smaller subset of the activities contained in any ANOP. More attention on process confirmation can then be paid to the areas of human intervention, which will improve the quality of execution of the entire operation, and the ease of confirmation. However, in order to automate any given DATAOP, the identifying information for data processing must be both clearly understood and contained within each DATAGRP as it is captured.

2. **Data Capture** - Data capture can consist of either manual transcription or automated capture of data. A manual capture step

generally proceeds initially by defining the DATAGRP appropriately with respect to sample and test. When capturing data automatically for presentation to a DATAOP, the system involved must also be able to find enough information to steer the DATAGRP through the appropriate data pathway within the DATAOP. At a minimum, the information available in the DATAGRP must include the sample number. Other information could be passed to allow the automation system to zero in on a particular test and parameter list.

3. **List Management** - Instruments are usually naive collectors of analytical information. For instance, many instruments do not deal with the issue of method (or instrument) detection limits, significant figure rules, and long parameter names. In these situations, test parameter lists act as controlled sources of extra information at various stages of data processing. At manual data capture, the lists act as a typing aid, serving up default information so that users are able to work more quickly. When instrument DATAGRPs are provided to the system, lists act as sources of information to complete data records. It is normal practice to have both manual and automatic data capture available for a given methodology. It is also possible to have alternate data capture scenarios available. This might occur in cases where individual lab departments sometimes capture data, and a report production department sometimes captures data. In both cases, lists are part of the generation of data; therefore, they must be controlled. The list name, date, and version must be captured within the automation system in all cases.

4. **Control of Data Identification** - Since multiple scenarios, data lists, and data processing paths are available within a DATAOP (see Figure 4), the automation system must be able to steer a captured DATAGRP properly to a given list, its information, and its path through a DATAOP. The identifying information must either be entered at the instrument workstation and presented in the data to the DATAOP automation system, or be added later by the analyst. Only when data are properly identified can

DATAOPs be automated. The control of identification information is a difficult task, since data collection workstations and manual data capture rely on humans to encode or choose identifying information.

5. **Standard Automation Procedure** - The next task then becomes properly specifying the data processing steps which constitute a DATAOP. When this is done, the natural result will be the ability of the quality assurance unit assigned to process confirmation to make use of the information in a structured way. For instance, this collection could be used to supplant some of the descriptive material in the method SOPs. We introduce in this work a new term, the Standard Automation Procedure (SAP), for this structured information collection. A Standard Automation Procedure is a dated, version-controlled, English language document which provides a description of the requirements of each data processing step for a given analytical operation for a given methodology. It is created programmatically from the DATAOP specification. The SAP is available to the quality assurance unit as a part of the documentation required by GLP. Since it is prepared from the actual DATAOP automation specification, it represents an accurate restatement of the process. An example is found in Appendix C.

6. **Control of Software** - An SAP must be implemented using an automation software system. The GLPs require control and documentation of software system. Therefore, one feature of the automation software system must be that the version of software which achieves any DATAOP step must be controlled and documented. Not only must an auditor be able to understand what version of a program was used to accomplish each data reduction step, but also to understand how the program works. The control of the operating software must also be addressed.

7. **Self-Documenting Software** - For control of software execution, each software module must be self-documenting. It must announce to the automation system its name, position in the file

system, and version for verification against the underlying records of the SAP.

8. **Data Auditability** - For the purposes of audit trail, each DATAOP processing step must mark up any data which it touches, move a copy of the previous data to the audit trail, and clearly identify itself as the author of the changes.

9. **Reporting Considerations** - Data reporting constitutes the release of data into an uncontrolled environment by the lab, since the lab has been cordoned off in our model from decisions based on lab data. An analogy with the sample chain of custody process suggests that this may not be entirely appropriate. The care taken during the entire lab process to ensure that the data generated represent the appropriate sample is not matched when data are released. The GLPs have no clear guidance on the subject of released data, except an implication that if an auditor starts with that data, the way in which it became available should be traceable and defendable. This seems to suggest that the execution of reports should be traceable in the same way in which processing of data is traceable. At the very least, copies of released data must be kept in paper or electronic format.

10. **Process Integrity vs. Data Integrity** - The GLPs strongly support control and confirmation of the integrity of data generating and manipulating processes. The broad base of data processing experience suggests that any process can be subverted, given enough reward. Therefore, the focus of confirmable automation should be to ensure that the system which delivers automation can be subverted only at high personal cost, and with a maximum of personal uncertainty. Unfortunately, these goals seem to be at odds with the normal goals of ease of use and development of understandable systems. The personal costs of inappropriate behavior is the subject of management practices. Raising the personal uncertainty that inappropriate behavior can work requires that, while the automation system is conceptually understandable and easy to use, the users are encouraged not to

understand the underpinnings or to interfere with the software. In addition, the software should appear to have many controls which can be used to detect and stop intervention.

The collection of requirements listed above are tall orders to follow. It is not surprising that data automation has not been totally successful in the GLP environment. However, when the requirements are well understood, capabilities usually follow. We need therefore to take a look at how a system with the characteristics above might be used within a system validation scenario.

APPLICATION OF THE DATAOP MODEL TO PROCESS CONFIRMATION AND SYSTEM VALIDATION

Software Robot™

One way to characterize the requirements of a DATAOP automation system is as a Software Robot. What is the definition of a traditional robot? Traditional robots, whether they build cars or automate CHEMOPs, do the following:

- Perform tasks
- Work without human intervention
- Repeat tasks again and again
- Work with instructions provided by the user
- Begin to work when instructed

A Software Robot automating DATAOPs does the same things, but has the additional task of documenting its actions and instructions within the data "objects" it works with and creates.

The traditional, hardware robot typically consists of three components:

- A mechanical, "doer" assembly
- A storage place for instructions
- A communication capability which allows instructions to be exchanged

In addition, it needs a way to recognize when it is time to go to work.

The Software Robot also contains these components. It must be capable of responding to a stimulus (usually the onset of a DATAGRP), and beginning its work. Its "working arm" must then execute DATAOP step tasks and make sure that all the proper data "bolts," pieces, and parts are used appropriately. The working arm must be able to select, calculate, check, test, and discriminate in order to handle data from a variety of methods and operations. The working arm must also "chisel" its signature in the data each time the data are touched, and create an audit trail of altered data. As with human data steps, any changes in the data must be accompanied by not only the identification of the robot entity which changes the data, but the date and time, and the reason for the change.

The Software Robot also works the same way traditional robots work, without human intervention. However, it performs its work repetitively as background computer tasks, somewhat less visible than its hardware counterpart.

Our robot also needs an instruction storage place, where it can find out what to do with the data objects which are presented to it. The instructions must be capable of specifying the information which allows the working arm of the Software Robot to execute all the data activities mentioned above. Since the instructions correspond exactly to the human-oriented SOP, changes in the instructions must be tracked and summarized and available in document format. It should come as no surprise that such a robot would be based on database software, using relational data tables to store instructions, data, and audit trails.

Lastly, the Software Robot requires the ability to communicate. Lab workers, data acquisition software, and automation system software itself must be able to send instructions and the signal to begin working.

When our robot works properly, the prospect of confirming a DATAOP seems much easier than before.

Confirmation Evidence

As we review the process confirmation goals for automation presented earlier, we note that many can be achieved by a properly operating Software Robot.

- **Isolation/Reduction of Human Intervention** - When a DATAOP is automated, our automaton performs its standardized actions on groups of data which have proper and complete identification. Humans are not involved in processing data unless they are specified as approvers of data. Their function is only to provide the identification necessary to instruct the robot how to process the data.

- **Strengthened Automation System Specification** - Specifications (or instructions) for the robot actions can be collected in a single place, then synthesized into a Standard Automation Procedure. The SAP can be used for GLP documentation in the same way that an SOP would be used.

- **Protection from Tinkering** - As with traditional robots, humans do not interfere with the workings of the Software Robot. While they may understand how to specify its actions, the lack of a complete understanding of its internal workings prevent personnel from actively altering data under its control.

- **Controlled, Self-Documenting Systems** - The SAP, the audit trails on specifications and data, the control over the robot's internal instructional software, and the robotic signatures on data records combine to create a an optimum situation for process confirmation. Lab personnel cannot make changes without telltale data signatures.

- **Simplified Confirmation** - The elimination of human intervention, SAP paperwork, mechanized audit trails, and complete log of operations allow DATAOP processes to be quickly confirmed, leaving more time for confirmation of human processes.

Validation

All these GLP confirmation benefits become available after automation
system software has become "validated." The concept of software
validation seems to fly in the face of current experiences with software,
requiring "completely" specified, tested, controlled, and buttoned-up
systems. How often has anyone experienced software, even at the operating system level, without bugs? Without unanticipated inputs which
prove problematic? Without workarounds? At the rate problems crop up
in Windows and DOS (the most used and most useful of operating systems), they should never appear in "validated" systems. Yet they do.

    Given the GLP requirement for system validation, there are some
conclusions to be drawn on how to go about it. First off, according to
regulators, automation systems may not be treated as "black boxes."
They must be proven almost from first "principles." The validation
process may not use confirmation by manual calculations alone as a
qualification. A validation effort must include a plan, use of boundary
condition checks, life cycle documentation, and more in the process.

Proof/Validation/Data Sets

Since the software robot works on groups of data which are properly
identified, the design of a validation effort should revolve around the
DATAGRP. Whether it is introduced to the robot by manual data entry
or electronic data capture, the DATAGRP has several characteristics
which lend themselves well to validation.

- **Identification -** First, DATAGRPs must be properly identified.
  Therefore, for each DATAOP, there must exist a definition of
  what constitutes proper identification. Examples of identification
  items might include sample numbers, analysis batch numbers,
  instrument IDs, test codes, analyst IDs, and more. If
  identification requirements are not met, the robot must know how
  to respond.

- **Instructions -** The robot must have instructions. If identification
  information does not point unequivocally to the proper set of

instructions, or points to multiple instructions, the robot must know how to fail graciously.

- **Bolts, Pieces and Parts** - DATAGRPs consist of records containing various pieces of data for a variety of entities (e.g., chemical compounds, weights, volumes, binary results). Some instructions work only on certain types of data records or samples. Validation requires proof that the robot works selectively on only the proper DATAGRPs.

- **Boundaries** - Boundary data checks are required by the GLP validation process. Boundaries may represent conditions such as those above, or may involve variations in the content of data records. Whenever calculations are involved, data boundary checks are required.

The use of the Software Robot during validation revolves around DATAGRP contents. The contents of a series of validation DATAGRPs can be tuned to exercise elements of the robots sensitivity to identification, instructions, bolts/pieces/parts, and boundary conditions. Since the robot leaves a complete log and marks up any data items which it touches, providing the appropriate evidence during validation becomes simply a matter of comparing or viewing the records.

CONCLUSION

We have developed in this work an approach to lab automation which has some applicability to GLP requirements. The framework began with defining some automation goals which would be useful during the monitoring or process confirmation activities required in the GLP regulations. Next we developed a generalized model of the lab which clearly separates the issues of automation into components of chemical, measurement, and data operations. An Economic Behavior Index then allowed us to make automation decisions on a scale which is responsive to the needs expressed by the GLPs. A focus on data operations (DATAOPs) then led us to the development of automation requirements which could satisfy the GLPs. This, in turn, suggested that a "Software

Robot" might be a technology which could be applied to DATAOP automation. As a final exercise, we developed the idea of the responses necessary in the Software Robot to allow process confirmation and validation.

Appendix A
Informing Process Inventory

ACTIVITY	FREQUENCY	TIME (min)	CFR REFERENCES
Requirements Definition			
Director	1	study	58.31a
Facilities	1	study	58.31e
Personnel	1	study	58.31e
Record education, etc.	1	person	58.29a
Record job description	20	study	58.29b
Materials	1	study	58.31e
Equipment	1	study	58.31e
Methods	1	study	58.31e
Records	1	study	58.35a

Appendix A
Informing Process Inventory
(Continued)

ACTIVITY	FREQUENCY		TIME (min)	CFR REFERENCES
Quality Assurance Unit				
Assign unit	1	study		58.31c
Educate personnel	20	study		58.31f
Update master schedule	1	week		58.35b1
Inspect study	1	week		58.35b3
Create status report	1	week		58.35b4
Monitor deviations	1	week		58.35b5
Review final report	1	study		58.35b6
Materials Testing				
Test control articles	1	man. batch		58.31d
Sample Scheduling				
Set up study schedule	10	study		58.35b1

Appendix A
Informing Process Inventory
(Continued)

ACTIVITY	FREQUENCY		TIME (min)	CFR REFERENCES
Sample Reception				
Login	1	sample	2	58.35b1
Assign tests	10	sample	2	58.35b1,58.63c,58.105a,58.113a
Label sample materials	10	sample	5	58.105c
Sample Storage				
Record location	1	test	3	
Workgroup Scheduling				
Create workgroup	1	batch	5	
Assign QC samples	1	batch	2	
Update workgroup	5	batch	2	
Instrument Operations				
Create and distribute SOPs	1	method	480	

Appendix A
Informing Process Inventory
(Continued)

ACTIVITY	FREQUENCY		TIME (min)	CFR REFERENCES
Clean	1	week	120	
Maintain	1	month	240	
Inspect	1	day	15	
Test	1	day	20	
Calibrate	1	shift	60	
Standardize	1	day	10	
Measurement	1.4	test	5	
Data Operations				
SOP creation	1	method	480	
SAP creation	1	source/method	120	
SOP/SAP documentation	3	method	30	
Capture data (manual)	1	measurement	20	

Appendix A
Informing Process Inventory
(Continued)

ACTIVITY	FREQUENCY		TIME (min)	CFR REFERENCES
Reduce data	5	measurement	20	
Forward data	1	measurement	10	
Report deviations	0.1	measurement	10	
Quality Control				
Inspect data	1	measurement	30	

Appendix B
Good Laboratory Practices
Economic Behavior Index (GLP EBI)[2]
for the Login Process

Applicability Criteria	Wt	1	2	3	4	5	6	7	8	9	10	Score	Wtd Score
Certainty of Audit Confirmation is unlikely to be stringent in this area(1). Confirmation will certainly be stringent (10).													
Audit Value GLPs refer to this process once or not at all (1). GLPs refer to this process many times (10).													
Completeness of Procedure Audit involves no interpretation (1). Audit of procedure requires interpretation (10).													

Appendix B
Good Laboratory Practice
Economic Behavior Index (GLP EBI)
for the Login Process
(Continued)

	Wt	1	2	3	4	5	6	7	8	9	10	Score	Wtd Score
Data Generation Criteria													
Volume Data for process has low volume(1). Data has high volume (10).													
Importance Data are not important to study (1). Data are produced at a high rate (10).													
Rate of Generation Data are produced at a low rate (1). Data are produced at a high rate (10).													
Number of Data Transformations Data are not transformed in this process (1). Data are transformed many times (10).													

215

Appendix B
Good Laboratory Practice
Economic Behavior Index (GLP EBI)
for the Login Process
(Continued)

	Wt	1	2	3	4	5	6	7	8	9	10	Score	Wtd Score
Number of Manual Steps Data are gathered or transformed by instrument (1). Data are gathered/transformed many times (10).													
Transfer of Data Data are used locally (1). Data are transferred to several locations for further use (10).													

Information Retrieval Criteria

	Wt	1	2	3	4	5	6	7	8	9	10	Score	Wtd Score
Nearness Information to be confirmed must be gathered from several places (1). Information is located in one physical place (10).													

Appendix B
Good Laboratory Practice
Economic Behavior Index (GLP EBI)
for the Login Process
(Continued)

	Wt	1	2	3	4	5	6	7	8	9	10	Score	Wtd Score
Difficulty of Integration Information is difficult to assemble or integrate into intelligible format (1). Information is easy to assemble (10).													
Number of Disciplines Successful confirmation requires multiple discipline interaction (1). Successful confirmation requires involvement of only 1 discipline (10).													
Number of People Confirmation requires many people (1). Confirmation requires a single person (10).													

Appendix B
Good Laboratory Practice
Economic Behavior Index (GLP EBI)
for the Login Process
(Continued)

	Wt	1	2	3	4	5	6	7	8	9	10	Score	Wtd Score
Frequency Confirmation is infrequent (1). Confirmation is frequent (10).													
Duration The confirmation process is short (1). The confirmation process lasts more than one day (10).													
Latency The confirmation data are stored on-site and are accessible (1). The confirmation data are off-site, archived, or inaccessible (10).													

Appendix C
Standard Automation Procedure
for Test PP-VOA

The test "PP-VOA," when performed in "SOLID" matrices, has two data capture/entry mechanisms, "GC/MS automated capture," and "Report Production manual data entry." These scenarios use the "MANUAL" and "GCMS FILE" data entry lists to control capture or entry, and to provide additional information such as MDLs, etc. The "GC/MS automated capture" scenario uses the "GCMS FILE" list, which is followed by a data operation chain called "VOASOLID." This chain reduces data in eight serial steps, with five additional non-serial operations possible (two of which are mandatory). The steps are as follows:

Serial Steps

Step	Process ID	Description	Automation	Type	Name Path
1	WGT/MOISTURE	Correction for weight	A	BACKGROUND	CALC/oracle/acsprod/seedpak3/idxl
2	BLANKQUAL	Set B qualifier	A	BACKGROUND	/oracle/acsprod/seedpak3/idxl

Non-Serial Steps

Process ID	Description	Mandatory	Type	Name Path
ANREV	Analyst review	Y	SCREEN FORUM	approval/oracle/acsprod/seedpak3/idxl

219

CALCULATIONS

Step 1: WGT/MOISTURE

The "WGT/MOISTURE" calculation, current version "C10," is designed to "Correct for sample weight and percent moisture." The "WGT/MOISTURE" calculation is performed on all sample types. It is applied to "REG" (Regular parameters) and "SURR" (Surrogate parameters) types of parameters only. No new information is inserted. "REG" and "SURR" type parameters are updated, but their parameter type is not changed. The "WGT/MOISTURE" calculation has no prior versions.

"WGT/MOISTURE" Information Requirements

For each parameter, find the "numvalue" column in "ng" units in the record. If it is not found, generate an error in this calculation and stop processing the data. Store the value in an intermediate variable called "ng."

For each parameter, find the "numvalue" column in "%" units of the "Percent Moisture" record in the same sample. if it is not found, generate an error in this calculation and stop processing the data. Store the value in an intermediate variable called "moisture."

For each parameter, find the "numvalue" column in "g" units of the "Sample Weight" record in the same sample. If it is not found, generate an error in this calculation and stop processing the data. Store the value in an intermediate variable called "wgt."

"WGT/MOISTURE" Calculation

The values "ng," "moisture" and "wgt" are used as follows in calculations:

Step	Result	Units	Formula
1	solids	%	(100-moisture)/100
2	numvalue	μg/kg	sigfigs((ng/wgt)/solids, sig_figs)
3	rdlvalue	μg/kg	sigfigs((mdlvalue*dilution)/solids, rdlsigfigs)

Step 2: BLANKQUAL

The "BLANKQUAL" calculation, current version "C92," is designed to "Set B qualifier hits in blank." The "BLANKQUAL" calculation is performed on "MS," "MSD" and "SAMP" sample types only. It is applied to "REG" (Regular parameters) types of parameters only. No calculation has 1 prior version, which expired on June 12, 1992.

"BLANKQUAL" Information Requirements

For the same parameter, find the "numvalue" column in "μg/kg" units in the "BLANK" sample. If it is not found, generate an error in this calculation and stop processing the data. Store the value in an intermediate variable called "blank."

For the same parameter, find the "rdlvalue" column in "μg/kg" units in the "BLANK" sample. If it is not found, generate an error in this calculation and stop processing the data. Store the value in an intermediate variable called "blankrdl."

"BLANKQUAL" Calculation

The values "blank" and "blankrdl" are used as follows in calculations:

Step	Result	Units	Formula
1	blankqual		decode(least(blank,blankrdl),blankrdl,"B","")
2	qual		translate(qual, "B", "")||blankqual

Non-Serial Step: ANREV

The "ANREV" (Analyst Review) is a "SCREEN FORM" step which *must* follow Step 6. Data are ready for reporting after "Analyst review."

Chapter 6

COMPUTER SYSTEMS VALIDATION

Sandy Weinberg, Ph.D.

Weinberg, Spelton & Sax, Inc., Boothwyn, Pennsylvania

INTRODUCTION

The Good Laboratory Practices (GLPs) provide valuable guidance for the organization and operation of a laboratory. Increasingly the real functionality of that laboratory is dependent upon the accuracy and reliability of a series of automated devices that control instrumentation, data management, archiving, interpretation, and reporting. If those automated systems fail to properly analyze, receive, store, interpret, summarize, or organize data, the integrity of the laboratory can be significantly compromised.

Unfortunately, a combination of poor quality control in the computer software industry, generally inadequate user controls, and the very complexity of the systems themselves have combined to erode confidence in the accuracy and reliability of computer systems. Horror stories proliferate; legitimate regulatory and managerial concerns are common; and the reputation of computerized systems is such that a presumption of

---

This chapter is adapted from *Validation* in the book *Good Automated Laboratory Practices*, Weinberg et al., Lewis Publishing, 1994.

confidence is no longer a norm. As in so many regulated areas of laboratory practices, skepticism prevails until support evidence is provided: proof of system control is now required.

That supporting proof of control is termed "validation." Validation, in this context, is the demonstration and proof of control of automated laboratory systems, including computerized instrumentation, laboratory information management systems (LIMS), data management systems, and sample control systems. Specific guidelines for the validation of laboratories have not been issued by the United States Food and Drug Administration, though an industry and agency consensus has provided a common understanding of the kinds of supporting evidence required. The United States Environmental Protection Agency (EPA) has codified that consensus in a draft guidance document titled "Good Automated Laboratory Practices," which serves as an excellent summary document of the current state of validation throughout the FDA and EPA regulated industries.

While the FDA has never endorsed the GALPs (largely for administrative reasons), and while the GALPs do not have the force of EPA regulation, they do provide valuable interpretive guidance and have been widely used by both investigators and field managers. The need for validation of GLP systems has been clearly established: the GALPs represent a practical, operational, functional definition of that validation proof. For a system to be compliant with specified GALP guidelines, a wide range of controls must be present. The GALPS summarize those tests and controls, with sufficient room for interpretation to meet the varying extingencies of wide ranging laboratory designs, purposes, and applications. But for a system to meet the GALP validation requirements, those controls must not only be present: they must be proven. The GALPS not only define appropriate procedures for validation, but also provide criteria for establishing the proof of those validation controls.

The skepticism underlying a demand for proof is not alien to either the scientist or the regulatory professional, yet somehow often emerges as a personal affront when representatives from the two camps interact. Perhaps this resentment emerges from history: the scientist has seen regulatory demands grow beyond reasonable levels, while the regulator has seen behind too many hollow facades claiming to be solid evidence.

In the computer automation field that skepticism may graduate into full scale cynicism. Technical complexities may exceed the expertise of both scientists and regulators, who have grown increasing uncomfortable with the jargon filled non-explanations of the computer professionals. Those computer professionals contribute to the atmosphere, too, with their resentments: their world has never previously had to surrender the shroud of authority for the ego-reducing discipline of double-check and confirmation. Finally, experience has created the need for supporting evidence: too many systems have failed in the past despite all the best promises of control and safeguard.

The result of this combination of history, reality, and attitude is a general regulatory dismissal of any presumption of system control. The "default situation," the unproven norm expectation, is that a system is not adequately controlled. Until firm evidence of that control is provided, an automated laboratory is considered to be without appropriate controls, and both the management and the data of that laboratory is suspect. The Good Automated Laboratory Practices define the controls that are appropriate: the validation portion of those GALPs define the proof that is necessary to establish compliance.

THE NATURE OF PROOF

Of the classic Aristotelian tripart definition of proof, only two techniques are relevant here. *Logos*, the logical component exemplified in laboratory systems by actual code and function tests, provides important confirmation of compliance. That logos can be verified, tested, and examined. It is the "hard" evidence on which a regulator, or manager, can rely. Included in this category would be actual logs, test records, original documents, and similar concrete findings.

Similarly *ethos*, the testimonial dependent upon the expertise and credibility of the witness, is critical. Evidence supplied by an impartial and credentialed observer may establish compliance with control Standard Operating Procedures, accuracy of documentary evidence, and suitability of code design. But whereas the accuracy of logos transcends its interpretation, ethos proof must be evaluated on the basis of its source. "Who said so?," "How does he or she know?," and "Why should he or she be trusted?" become the key questions. It is upon the importance of ethos that the important issue of independent, "quality assurance," confirmatory investigation lies. Most ethos testimony takes the form of reports, observational records, and certifications.

But *pathos*, the passionate belief of faith, does not apply. A programmer may "know" his code is sound; a manager may be confident her workers are well trained; a supervisor may be convinced the system is reliable. These beliefs are critical, and are not to be disparaged: effective control would not be possible with ultimate reliance upon such well placed and reality tested faith. But pathos is non-evidentiary: it can not be evaluated independently, and falls beyond the realm of science or of regulation. Validation must rely on proof: confidence may point the path toward obtaining such evidence, but is not a substitute for it.

While this may seem a self-evident conclusion, the subtlety of pathos is pervasive. How do we know the system is functioning? The self-diagnostics tell us so. And how do we know those diagnostics are accurate? Ultimately, we must rely upon faith, but that faith is not acceptable regulatory evidence regardless of the passion behind it. Effective evidence, though, buttresses that faith with varying levels of confirmatory evidence: the oscilloscope is calibrated; the testing tool is independently tested; the observer passes the test of independence. Without such checks, data generated by systems can not be consistently trusted in any scientific sense, and an endless spiral of insupportable claims are left devoid of control.

In the earliest days of computer systems highly inflated estimates of the power, potential, and accuracy of systems created a strong pathos of proof. The "computer says so" became the rallying cry and defense of billing agents, government clerks, and bureaucrats the world over. But as stories of enormous and humorous computer errors flooded popular culture in later years, a "computer error" became as common a punchline as the "check is in the mail:" computer professionals fell from god status to a reputation probably far below the reasonable norm of accurate and reliable system function. The result was, and is, an appropriate demand for controls, even as most reviews demonstrate that those controls are preventive rather than corrective of real problems.

In the appropriately skeptical world of interaction between laboratory scientists and the regulators who must rely upon their conclusions, proof of control must flow from the evidence of logos and ethos. In effect a past history of poorly designed, implemented, and controlled systems has destroyed any pathos to which computer professionals may have otherwise been entitled.

VALIDATION EVIDENCE

Exactly what kind of evidence of validation is required? How much evidence is sufficient to establish clear control? These questions can be answered through an examination of two dimensions. Validation evidence falls into six broad **issue** categories, further defined by two cross matrices of **risk** and **application**. Before defining these two cross matrix dimensions, though, a detailed description of the issue categories will be helpful.

Evidence of Design Control

Evaluation of any automated laboratory system ultimately involves an assessment of the appropriateness of that system to the job for which it was intended. If the system adequately performs its intended (or assigned) task, it is useful. Regardless of elegance and accuracy, the system is useless if it does not meet the parameters of its application. A bar code system may be intended for tracking samples. No matter how well the software functions, that bar code system is worthless if it does not assign unique numbers and hence fails to allow unambiguous tracking. While such a match seems a self-evident requirement, incompletely considered or changing needs often result in systems being used in situations inappropriate to their design.

The key to matching design with system is an effective and up-to-date needs analysis. This process of clearly defining and documenting purpose not only serves to assist in the process of selecting or building systems, but also serves as a post facto template for managerial and regulatory evaluation of a system. Without a clear statement of exactly what a system is intended to accomplish, it is impossible to determine whether or not that (non) goal is met.

Formal needs analysis approaches often use sophisticated survey and data flow analytical tools to produce a detailed request for proposal from vendors or comparison model for purchase evaluation. Even the least formal needs analysis must provide three kinds of critical information.

First, the outputs or end results of the system must be clearly defined. In many environments both the format and content of that output is critical. For example, a specific EPA water testing project may require reporting of lead values, and may require that those values be printed in a specific location block on a specified form. All outputs should be unambiguously defined, generally through modelling the actual reports or screens that will be required.

Second, the sources of those output elements must be specified. Some outputs are user (or related system) entered. For example, a Laboratory Information Management System (LIMS) may receive the water lead levels from a chromatography system. Other outputs may be derived from entered data, perhaps through reformatting the reported lead levels. Finally, some data may be system generated, perhaps through comparing the received lead level to the average of all other samples and making the determination of whether or not to label a given sample outside of norms.

Finally, the dimensions or ranges of all variables (the outputs and their sources) must be specified. If a system is intended to handle 500 samples per day and can only accommodate 200, it is appropriately criticized. If lead levels are required to three decimal points, a system limited to two decimals is inappropriate. The range of variables is an important specification of system user needs.

These three kinds of information, along with other supporting documentation, must be provided as evidence (logos) of the system design. The review that documentation, assuring its appropriateness, thoroughness, and the degree to which it was followed, provides the additional evidentiary support (ethos) for the system validation.

Evidence of Functional Control

When a system is first installed or utilized, it should be subject to detailed and thorough user testing, including use in parallel to previous systems for a specified period of time. Only when the existing system and new system have produced consistently matching results, or some other comparison process has been used, should the new system be considered acceptable. Even so-called "standardized" software should be subject to this rigor of testing, since unique application or configuration parameters may effect the functionality of the system.

Post acceptance periodic retesting is prudent, and retesting after modification, crash, or problem is all but mandatory. Most of these acceptance and confirmatory tests are user designed and implemented, however, providing only limited value as confirmatory evidence. While the tests themselves stand as evidence, the review and analysis of those tests, and review of the test designs, requires user, developer, and vendor independence for the establishment of credibility.

The validation process provides that ethos by reviewing all test protocols and scripts for thoroughness, appropriateness, and applicability; by replicating a sample of tests to confirm functionality; and independently analyzing the results to arrive at conclusions of acceptable significance levels.

The user tests and validation tests fall into two overlapping divisions: within range (normal function) and out-of-range (stress or challenge) tests. The normal tests evaluate system functionality in expected use. The challenge tests examine performance when parameters of variable, range, and dimension are violated. Ideally, norm tests should show results matching to independent confirmatory sources. Challenge tests should show system rejection of inappropriate data, and system maintenance of data base integrity despite stresses. Because of the potentiality for data corruption, challenge tests particularly should be performed on non-live (library or test) systems.

Evidence of Operational Control

If systems are inappropriately used, the results of those systems are questionable at best. Validation review of a system must include an analysis of proper use, and an evaluation of the degree to which normal user behavior falls within those proper use norms.

Norms are established through the development of Standard Operating Procedures (SOPs), Technical Operating Procedures (TOPs), and working guides (such as help screens and manuals). Those procedures are communicated to users through a combination of memo, manual, training, and support.

The formal Standard Operating Procedures shall be discussed in further detail in the next section (Managerial Controls), since they represent the high level policy decisions of laboratory and system managers. The implementation of those policies is generally specified in the TOPs that detail user activities.

Some laboratories may combine SOPs and TOPs in single documents, consisting of a policy and the detailed directions for carrying out that policy. Such a documentary combination is acceptable, but is not recommended, since it requires a lengthy and unnecessarily complex review process for even the most minor modifications. For example, an SOP may call for safe storage of back up system tapes; a TOP may specify the room to be used for that storage, and the inventory procedures for maintaining that room. Should the number of tapes necessitate moving to a second or larger storage room, the TOP can be amended efficiently. If the same change is required within an SOP, a much more complex managerial review process may be required.

The documentation of procedures to be followed, including training outlines and manuals, are an important part of the validation evidence. Accompanying that documentation should be an expert review for appropriateness; and a confirmatory observation to determine the degree to which those documented procedures reflect the realities of the laboratory.

Evidence of Managerial Control

In small laboratories the lines of control are simple and straight forward: often the manager and lab technician may be the same person. But as laboratories grow in size and complexity the potential increases for a communication problem between the manager of that laboratory and the people involved in basic laboratory activities.

In the regulatory world the manager of a laboratory has a unique role: he or she assumes formal responsibility for the activities and results of that lab. That responsibility is predicated upon the assumption of clear and unambiguous two-way communication: the manager has clearly provided instructions to the lab technician, and the technician has provided effective feedback concerning those directions to the manager. These control issues are significant regardless of the degree of automation in the laboratory. If the laboratory is computerized; however, the control becomes more complex, since the computer in effect becomes an intermediary in the chain of communication: the manager programs or causes to be programmed the recipes and data bases for the various tests, which in turn provide instruction to the lab technician. Similarly, the technician enters the data into the system, and the computer provides reports and summaries that provide the control feedback to the manager. With the computer in this intermediary position, managerial control of the system becomes a critical issue in controlling the laboratory and assuming regulatory responsibility for activities and results.

Managerial control is established and documented through a series of Standard Operating Procedures. These SOPs are system design, use, and control policy statements. They summarize procedures of system security, disaster recovery, normal use, data archive and back-up, error response, documentation, testing, and other important aspects of control.

Each Standard Operating Procedure must meet three tests in order to demonstrate control. First, the SOP must be <u>appropriate</u>: that is, a review by management must establish responsibility for the procedures specified, presumably with the evidence of a signature (or, in the emerging future, an electronic equivalent). Second, the SOP must be <u>timely</u>. That is, the review must be dated, generally within the past twelve months, confirming that the procedure is still appropriate to the situation. Most organizations provide for an annual re-review of all SOPs, including those related to system control. Finally, the SOP must be <u>available</u>: all pages must be clearly in the hands of all appropriate personnel, and only those pages appropriate should be in distribution. This requirement presumes some sort of clear recall and control mechanism, some paging control, and some method of SOP storage or posting.

Evidence of Data Integrity

Once data has been appropriately and accurately entered in the system, processed, and stored, it is presumably available for later comparison, analysis, or combination. But that presumption is based on confidence that the system does not in any way corrupt or modify the data. Validation requires evidence of continued data integrity.

Four areas of potential threat to data integrity need to be addressed, presumably through a combination of tests, policies (SOPs), and design features. First, and of greatest regulatory interest though probably not

very high in reality of threat, is the question of data security. News
stories of "hacker" and "virus" attacks of systems have created a high
awareness of the potential dangers of malicious or unprincipled attempts
to enter a data base. Effective protection from security threats has
become an important focus of data integrity proof. These protections
most often take the form of system locks (physical locks, passwords,
software keys, etc.), system isolation (controlled modem access, physical
site protection, etc.), and violation trails (logs, audit trails, etc.). In
balanced and reasonable proportion these security protections can prevent
or detect any threat to data integrity.

Interestingly, too much security can have the undesired effect of
reducing protection. If controls are too rigid, making normal productivity
difficult, workers have a tendency to develop techniques for circumvent-
ing security measures. Complex electronic key doors are wedged open.
Passwords are recorded on desk calendars. Systems are not turned off
when unattended to avoid complex log-in procedures. In developing
security controls a balance with appropriate access must be considered.

Second, disaster situations represent real and potential threats to data
integrity. Evidence of appropriate preventive action and recovery
strategies must be presented, generally in the form of a Disaster Recovery
Plan with an annual practice drill. The Disaster Recovery Plan is usually
organized around likely problems (flood from broken pipes, fire, electrical
failure, etc.) and includes appropriate notifications, substitute activities,
and recovery actions. The Disaster Recovery Plan generally interacts
with system back up, recovery, and archive SOPs.

Third, problems of data loss in transmission must be addressed, with
evidence of prevention and control strategies. These strategies generally
relate to the transmission channels, if any, in effect. The use of
bisynchronous channels, bit checking procedures, and check digits
commonly provide evidence of transmission control.

Finally, data threats related to environmental conditions have
generated a great deal of publicity (though in reality are probably very
minor). Laboratories located on radon spurs, or located in or adjacent to
nuclear facilities, need to be concerned about magnetic and other
radiation that may corrupt stored data. An inspection and data
reconstruction test generally provides sufficient control proof.

Evidence of System Reliability

All of the areas of proof described above provide evidence concerning the current operations of the computer systems in place. But can those same controls be expected to continue to function over time? Certainly a trend of control provides some presumption, and annual SOP review procedures provide a degree of assurance. But the most significant evidence of system reliability lies internal to the software, and is documented only through a review of that source code itself.

Future confidence is based upon the organization of the code, the accuracy of the formulae and algorithms incorporated, and the "elegance" or simplicity of the code. These elements are the focus of the code review.

Poorly organized "spaghetti" code, filled with convoluted pathways that jump back and forth within the code stream, make continued support difficult and create an environment in which future changes are likely to cause unanticipated problems. Alternately, well organized code allows efficient maintenance with appropriate tracing and variable tracking.

Consistent and proper operation of any software system is dependent upon the decision and action formulae or algorithms included in the code. With a poorly designed algorithm, interim problems may not be obvious in testing, but may cause significant difficulties over time. Similarly, improper formulae may work properly with some data sets, but may malfunction with unusual or "outlier" data points. Examination and confirmation of appropriate formulae is a critical part of any source code review.

Finally, many complex software programs are modified or evolved from other programs. The result may be convoluted dead end pathways, non-functioning "dead code", and inefficient module looping structures. Examination of code to determine the elegance or simplicity that avoids these non-parsimonious problems, provides an important element in the evidence supporting continued reliability.

The proof in support of reliability is a combination of the logos of the actual code (or reviewed subsection samples), and the credible report examining the elements described above. Here particularly the expertise of the examiner, establishing the thoroughness and soundness of judgement concerning efficiency and reliability of the code, is of particular importance.

THE INFLUENCE OF RISK AND APPLICATION CONSIDERATIONS

The six proof areas described above identify the topics for which evidence must be gathered, but what evidentiary weight is required? How much testing is sufficient? How detailed must a review be? How large a sample of code should be analyzed? When is "enough" enough? The responses to these questions evolve from art rather than science: no absolute definitions are available, no inflexible yardsticks exist. But two parameters provide important guidelines that can be used to generate defensible responses for the vast variety of situations to which the concept of validation applies: "risk," and "application."

"Risk" refers to the danger resulting from a system-related error. In a blood processing center, for example, the computer may calculate the appropriate disposition label for a bag of donated blood. If algorithms are incorrect or data is scrambled, a dangerous bag of hepatitis or AIDS positive blood may be incorrectly identified as safe for human use. In such high hazard situations, testing of the computer system must be comprehensive, thorough, and redundant.

At the opposite end of the spectrum, consider a computer system used to track inert material in a warehouse. Errors in the system may inconvenience the production schedule, but have little or no chance of causing real harm. Even a complete misidentification will be quickly corrected in a QA test of final product. For such a system some validation is still necessary, but a higher tolerance could be used: smaller samples, less frequent rechecks, and broader testing parameters.

The fluid nature of this broad range of risks, and the non-specific relationship to the depth or extent of gathering proof, argues further for the expert nature of the process. Only the combination of experience and training that qualifies a true expert will allow consistently appropriate decisions with such inconsistent and murky criteria. The alternative, defining all systems in terms of the most rigid requirements, is an expensive and unnecessarily burdensome alternative.

A further honing of proof quantification comes from the concept of "application." Software can be broadly defined as "standard" (widely used, as with an operating system); "customized" (the multiple site development of a shell defined and written from a standard program, such as the development of Statistical Analysis System (SAS) in a C-based language); and "unique", software written specifically for a single user or site (perhaps, using the same example, the specific protocols written in SAS for use in a specific study).

In principle, the experience of other users with a broadly based system can mitigate the responsibility of any single user. In practice, the need to invest effort in more than a cursory testing of software is eliminated in the standardized packages (except in high risk situations!); for customized software, adjustments in the sample size, depth of analysis, and other factors may be appropriate. As in all safety situations, default should be to the high level: that is, if unsure of the reliability or standardized nature, increase the validation effort. High risk situations will always argue for increased vigilance, regardless of the number of sites sharing an application.

ERROR LOGS AND PROBLEM REPORTING

The on-going use and enhancement of a particular application system on a given hardware platform and the installation of additional systems will entail problems and/or failures. In the regulated environment, it is not sufficient to observe that "stuff happens" and continuing processing. There is a special requirement for reporting, classifying, responding to and resolving problems. This can be the operational companion to rigorous designs and coding standards. Even rigorous system development practices, which carefully document and control design changes, can de defeated by inadequate trouble reporting procedures. End-users who have access to coding or report generation tools can take it upon themselves to modify and/or enhance what they perceive as an inadequate system.

The discipline of the regulated laboratory requires the equivalent of a notebook or log, physical or electronic, that will record problems. The recording by itself, however, is not sufficient evidence of control. The tracking and resolution of these problems both demonstrates that active measures are being taken to control the system. These entries, linked to activities required to enhance/update the system, provides evidence that the required activities are actually being performed. In addition, these provide an outside auditor with another frame of reference for seeking and reviewing evidence of control.

THE VALIDATION REPORT

The six areas of proof previously discussed also provide a comprehensive package of evidence in support of the Good Automated Laboratory Practices. Each area is supported with specific documentary evidence such as test results, SOPs, manuals, and code; and with testimonial evidence in the form of evaluations, interpretations, and summary reports.

Since the report is in itself a "snapshot" picture of compliance at a given period of time, it should be updated periodically. A complete revalidation is not necessary, but many sites find that an annual review of the validation report is helpful. Occasional specific events, such as upgrades of programs or replacement of hardware, may trigger partial or complete retesting. Finally, complex systems tend to evolve, so a review to confirm that version control procedures are appropriately followed is recommended on a regular (at least annual) basis.

The report should also establish the credentials of the validating team.

CREDENTIALS

Since the most significant portion of validation evidence rests upon ethos proof, the credentials of the validators are of utmost importance. The credibility of their collective testimony relies upon their expertise, and the objectivity of their conclusions. That expertise is a matter of education and training, experience, and access to appropriate tools and technics. The objectivity that underlies their credibility, however, is a matter largely of organizational structure.

In any organization a series of reporting relationships define interactions between persons and groups. Those interactions include basic communications, but encompass more complex interactions including employment and evaluation issues. In the classic Quality Assurance model a separate and distinct unit, outside the normal departmental reporting relationships, is used to audit function and activity. The independence of this QA team, free from personal evaluations and budgetary decisions, assures an objectivity of examination. Validation follows the same line of approach. To maximize the credibility of the validation, and the value of the testimony provided, validators should be independent of normal lines of authority. Either operating as outside consultants, as an autonomous quality assurance unit without direct reporting lines to the laboratory or lab management, or through some other mechanism, independence must be assured and proven.

Defining appropriate expertise is even more complex. One credentialing group associated with Weinberg, Spelton & Sax offers a "Certified Validation Professional" credential for individuals who demonstrate a combination of academic and non-academic training in regulation, statistics, systems theory, and laboratory skills; and experience in auditing, testing, and validation. Using such a certifying agency or working individually, the credentials of the validator or validators should be established and provided as an important part of the validation report.

CONCLUSION

Validation establishes the credibility of laboratory data and automated procedures. Without a credible validation review it is certainly possible to follow the GALP guidelines or equivalent industry consensus: validation provides the proof that those guidelines are incorporated in daily and on-going activities. The GALPs serve two important proposes: they establish the agenda for managing an automated laboratory, and they provide a framework for regulatory review of that laboratory management. Without validation the first purpose can be effectively met: managers can check results, document activities, organize controls, and develop security precautions without any independent check upon their activities. But demonstrating compliance requires validation, for it represents the proof that that agenda is followed.

Could regulators conduct their own audits, not depending upon validation by laboratories? In theory that strategy could be successful, but two problems stand in the way. First, resources, including time and expertise, permit only a very cursory spot check on compliance. Those limited resources are much better spent in the review of comprehensive validation reports rather than in conducting very limited tests of system performance and compliance.

Perhaps more importantly, though, is a fundamental philosophical limitation. Is a laboratory manager willing to be so dependent upon a computer system that the only confirmatory check upon automated data is provided by a regulatory inspection? That acceptance would seem to be a real limitation on the kind of control the GALPs, and indeed the Good Laboratory Practices themselves, are designed to encourage. Rather than accept blindly system generated results, validation represents prudent checking on system performance.

As a result, validation represents a prudent, cost effective, and efficient way of assuring regulatory acceptance and of assuring internal control of automated laboratories and the system upon which they rely.

Chapter 7

THE FDA's GLP INSPECTION PROGRAM

George W. James, Ph.D.

U.S. Food and Drug Administration, Rockville, Maryland

PURPOSE AND OBJECTIVES OF GLP REGULATIONS

The Federal Food, Drug and Cosmetic Act (Act) is enforced by the Food and Drug Administration (FDA) to assure that all regulated products, including food and color additives, animal food additives, human and veterinary drugs, medical devices for human use, biological products and electronic products, are safe and effective for their intended use or uses. FDA accomplishes this responsibility regarding safety by suggesting the type and extent of testing that is required, by reviewing new product applications to determine whether the contemporary scientific standards of safety have been met; and, in certain circumstances, by carrying out independent scientific studies to confirm the results submitted by product sponsors. Further to this end, FDA requires that all nonclinical toxicity studies be conducted under conditions that assure that the resultant final report is suitable for informed regulatory decision making. The Agency believes that this requirement can be met if the toxicology laboratory is operating in accord with universally accepted principles of Good Laboratory Practices (GLPs).

239

Each of the five Centers of the FDA, The Center for Drug Evaluation and Research; The Center for Biologics Evaluation and Research; The Center for Food Safety and Applied Nutrition; The Center for Devices and Radiological Health; and, The Center for Veterinary Medicine has a special unit that oversees compliance with the Good Laboratory Practice (GLP) regulations; Title 21, *Code of Federal Regulations*, Part 58. The GLP activities of these Centers are coordinated in the Office of the Associate Commissioner for Regulatory Affairs as part of the FDA's Bioresearch Monitoring Program. Both nonclinical and clinical research is included in this program.

FDA has developed a toxicology laboratory monitoring program to conduct vigorous inspections intended to foster and to verify adherence to the principles of the GLPs. The objectives of this program are: to inspect nonclinical laboratories engaging in studies that are intended to support applications for research or marketing permits for regulated products to determine the degree of their compliance with the GLP regulations; to audit ongoing and completed nonclinical toxicity studies to verify their integrity and validity; and, to initiate appropriate corrective actions when GLP violations are encountered. The details of the program are contained in the FDA Compliance Program 7348.808.

TYPES OF GLP INSPECTIONS

There are two types of GLP inspections. The first is the routine inspection, a periodic evaluation of a laboratory's compliance with the GLP regulations. To facilitate scheduling of routine inspections, the Agency maintains a list of nonclinical testing laboratories actively engaged in the toxicity testing of regulated products. These laboratories are inspected for GLP compliance at least once every two years. FDA reviews the list for scheduling inspection assignments and the list is updated when FDA becomes aware of new facilities.

In preparing for a routine inspection, it is necessary to select toxicology studies for audit that are as representative as possible of the laboratory's current operations. This is done either by the assigning Center's GLP unit prior to the inspection or by the field investigator at the laboratory site. The selection, when made by the field investigator, is drawn from the firm's GLP master schedule sheet.

The GLP master schedule must list all of the studies conducted at the laboratory that are subject to the GLP regulations. This master schedule, indexed by the test article, must describe the test system, the nature of the study, the date the study was initiated, the current status of each study, the identity of the sponsor and the name of the study director. The field investigator may, using the GLP master schedule sheet, exercise the option to select a study or studies that another of the other FDA Centers are required to evaluate for scientific content, rather than the studies designated by the Center assigning the inspection. For example, if a testing facility to be inspected does not have an ongoing drug study, then a food additive, a veterinary drug, a medical device or radiation emitting product safety study, could be selected for audit. In such instances, the GLP staff for the assigning Center forwards the information concerning the audited study to the appropriate Center's GLP component for review and follow-up action.

The second type of GLP inspection is the directed, or *for cause,* inspection. The directed investigation is more complicated by its nature than the routine and is less frequently performed in the GLP program. These constitute only about 20% of the GLP investigations completed since the regulations were invoked.

Directed inspections are assigned for one or more of the following reasons:

1. To determine if appropriate actions have been taken by a firm to correct serious GLP deficiencies noted in a routine inspection. This is normally done six months after the FDA receives the firm's assertions that corrections have been made.

2. To resolve concerns raised in the preclearance review of final study reports submitted to research or marketing permits, such as an IND or a NDA.

3. To validate critical studies, such as long term and reproduction toxicity studies, submitted to INDs or NDAs. These studies are selected at each Center from master schedules collected in the course of previous GLP inspections or from reviews prepared by pharmacologist responsible for evaluating applications for research and marketing permits.

4. To verify validations performed by a third party for the sponsor.

5. To investigate seemingly questionable circumstances brought to the FDA's attention by other sources such as the news media, other operating firms or laboratories or disgruntled employees.

OPERATIONAL ASPECTS

Logistically, the inspection is a field operation. One of 22 FDA district offices located throughout the 50 states and Puerto Rico will assign the field investigators to perform the inspections. Usually, investigators perform routine investigations alone.

Headquarters' personnel, such as representatives of the Office of Regional Operations (ORO) and the Office of Enforcement, pharmacologists of the GLP staff of the assigning Center and, on occasions, scientists from the reviewing divisions may be asked by the assigning Center to participate in the GLP investigations.

ORO acts as a contact for the arrangements involving headquarters' participation in the inspection. The field investigator, designated as the team leader, has the responsibility for the conduct of the inspection and the preparation of the inspection report, known officially as the Establishment Inspection Report (EIR). The lead investigator begins preparation by contacting any headquarters' personnel identified to participate in the assignment in order to make the necessary arrangements for coordinating the inspection.

Another important preliminary to the inspection is the preinspection conference that is usually arranged to include all members of the inspection team as well as any other field and headquarters' specialists judged appropriate by the FDA Center assigning the inspection.

NOTIFICATION OF INSPECTION

Prior to 1991, after the inspection team had been formed, the next step was for the district office to notify the laboratory of the pending inspection by telephone, about one to two weeks prior to the inspection.

Since 1991, however, laboratories to be inspected are not given advance notice.

AUTHORITY TO INSPECT

FDA can only enforce inspection of laboratories that perform tests on foods, drugs, new animal drugs or medical device products. Should a laboratory assumed to be doing nonclinical toxicity studies refuse to permit inspection, the laboratory will be advised by the FDA investigator that it is the policy of the Agency not to accept studies submitted in support of any research or marketing permit if the Agency does not have inspectional information regarding the GLP compliance status of the firm. Even partial refusals, such as refusal to permit access to copying of the master schedule sheet and its code sheets, Standard Operating Procedures (SOPs), and other documents pertaining to the inspection, are treated the same as a total refusal to permit inspection.

ELEMENTS OF A SURVEILLANCE INSPECTION

The first part of the surveillance inspection covers organization and personnel, which are addressed in Parts 58.29 through 58.35. Investigators must determine whether or not the facility has an adequate number of qualified personnel to perform the types and numbers of nonclinical laboratory studies that it has been or is, performing. FDA investigators describe in the EIRs the organizational structure and competency of the laboratory. To do this, FDA obtains an organizational chart and the summaries of training and experience of the managers, study directors and other appropriate supervisory personnel. If personnel are involved in studies in a location other than that of the inspected facility, the sites and the personnel so involved must be identified. In fact, if there is a need for an inspection of the outside contract facility, this must be specifically noted in the EIR. As part of the organization and personnel evaluation, programs used to increase training and qualifications of personnel through inhouse and outside programs must be included in the EIR. As part of this evaluation, FDA must identify, through review of the facility personnel SOPs, how the facility recognizes and deals with health problems of the employees, especially those problems that may affect the quality and integrity of studies being performed by that individual.

The Quality Assurance Unit (QAU), whose duties are described in Part 58.35, presents a special challenge to the FDA investigators. By evaluating QAU activities, the Agency is able to assess the mechanisms by which the facility management assures itself that the nonclinical laboratory studies are conducted in a manner that will assure the quality and integrity of the data generated in the laboratory. This is most commonly accomplished by obtaining a list of the QAU personnel and the written procedures for QAU study audits and in-process inspections. The master schedule is also an important tool in the assessment of QAU activities. With it, the investigator can determine whether or not the QAU adequately maintains master schedule sheets and protocols with any subsequent changes or amendments. FDA investigators should always obtain copies of master schedule sheets dating from the last GLP inspection or covering at least the last two years. Sometimes, the master schedules are voluminous and the investigators may take only representative pages for headquarters' review. Also of interest are the methods by which the QAU schedules and conducts audits. Investigators determine how the QAU retains records and to whom the QAU reports its findings. The records of QAU findings and the records of corrective actions recommended by the QAU and acted upon by management are normally exempt from routine FDA inspection. One exception to the FDA policy of not requiring access to QAU findings and corrective actions recommended and taken, is that the Agency may seek to obtain these reports during a litigation under procedural rules as applicable for otherwise confidential documents.

Parts 58.41 through 58.51 cover the physical facilities of the laboratory. The inspector must determine whether or not the facilities are of adequate size and design for completed or in process studies. The physical parameters and systems of the facilities as they are used to accommodate the various operations employed in the GLP studies are examined. Investigators also deal explicitly with the environmental control and monitoring procedures for critical areas especially the rooms used for animal housing, the test article storage areas and the laboratory areas where biohazardous material is handled. The procedures and methods for cleaning equipment and areas critical to study conduct as well as the current status of cleanliness are also closely examined. It must be determined that separate areas are maintained in rooms where two or more functions requiring separation are performed and how that separation is controlled and maintained. The facility inspection must examine the adequacy of pest control procedures, especially in storage and animal housing areas. This is important because residues and improper use of insecticides and pesticides have been known to impact on the result of GLP studies.

As would be expected, equipment is also of considerable interest to the FDA investigators. This is covered by Parts 58.61 and 58.63 of the GLP regulations. It must be determined whether or not the facility has sufficient equipment to perform the operations, which are specified in the protocols and that such equipment is maintained and operated in a manner which ensures valid results. This is done by examining the general condition, cleanliness and ease of maintenance of the equipment in the various parts of the laboratory. Also, it must be determined that the equipment is located where it is to be used, and, if necessary, located in a controlled environment. For representative pieces of equipment, the investigators check for SOPs, maintenance schedules and logs, standardization/calibration procedures, and, finally, it must determined if standards for calibration and/or standardization are available. Investigators must be aware of any equipment deficiencies that might result in contamination of test articles, uncontrolled stress to the test system and/or erroneous test results. Investigators also learn if the same equipment is used to mix test and control articles, and, if so, whether the procedures are adequate to prevent cross contamination.

FDA investigators must give particularly close attention to Parts 58.81 and 58.83, which address the testing facility's standard operating procedures (SOPs). They must judge whether the studies are being conducted in conformance with these SOPs and in a manner designed to assure the quality and integrity of the data. To accomplish this, they obtain copies of the index and representative samples of all of the laboratory's written SOPs. Furthermore, these SOPs must be available at the locations where they are to be used. All SOPs and any changes to the SOPs must be appropriately authorized and dated and historical files of SOPs must be maintained. The procedures for familiarizing employees with SOPs must also be reviewed.

Part 58.90 of the regulations deals with animal care. Animal care and housing must be adequate to preclude stress and uncontrolled influences that could alter the response of test systems to test articles. The personnel responsible for receiving and examining animals and the animal care procedures including any routine treatments, such as vaccination and deworming, are evaluated. Further, FDA examines the criteria used to determine when and for how long animals should be kept in quarantine. Relative to this, GLPs used to separate species and the methods used in handling or isolating diseased animals are examined. At the same time, the method of uniquely identifying newly received animals can be determined.

One of the most important aspects of any nonclinical laboratory study is the preparation and presentation of test and control articles to the test system or test animal. Parts 58.105 to 58.113 of the regulations address this. FDA reviews the procedures used to ensure that the identity and the dose of test articles administered to the test systems is known and is as specified in the study protocol. In the course of assessing this, the investigators evaluate the methods used in the acquisition, receipt and storage of test articles and the means used to prevent deterioration and contamination must be evaluated. The identification, homogeneity, potency and stability of the test articles and the means used to determine these parameters are also closely examined. The methods used to ensure test article integrity and accountability and for retaining and retesting reserve samples of test and control articles must also be evaluated. The aspects of diet mixing that should be observed include: the frequency and methods used to determine uniformity and accuracy of mixing and the stability of test article carrier mixture; the labeling, storage and distribution; the disposal of the test article carrier mixture; and the identification and specification of carriers and/or feeds.

Parts 58.120 and 58.130 address the protocol and the conduct of the nonclinical laboratory study. FDA judges whether or not the facility's protocol is generated, approved, changed or revised in conformance with the GLPs. The overall test system monitoring, specimen labeling and data collection procedures must be described for the EIR.

The final portion of the GLP surveillance inspection includes examination of records and reports as described under Parts 58.185, 58.190 and 58.195. To accomplish this, FDA assesses the facility's ability to store and retrieve study data, reports, specimens, etc., in a manner which maximizes their integrity and utility. This must include an overview of how the firm maintains materials such as the raw data and the various specimens that are developed in the course of the study. The investigators must become familiar with the facility's archives regarding their location and the accessibility. The individuals responsible for the archives must be identified and FDA must learn whether or not the archivec is indexed, and if the materials and records that have been transferred and stored elsewhere are appropriately identified. Furthermore, the procedures for adding or removing materials from the archives must be examined and individual test systems are selected randomly to determine that all raw data, specimens and documents, have been retained as required.

The examination of records and reports usually concludes the GLP surveillance inspection of a facility, although there may be extenuating circumstances that will prolong the investigation and require closer review of a given area.

STUDY AUDIT

The most important aspect is the audit of completed or ongoing studies. This is particularly true of directed inspections which essentially is an audit of studies. There are basically two reasons for conducting a study audit during a surveillance inspection. First, there is a need to determine whether or not compliance with the GLP principles by the nonclinical laboratory has resulted in valid studies. Second, to determine if a study, or studies, either critical or suspect, have indeed been appropriately conducted.

There are ten prime areas of a nonclinical toxicity study that must be examined:

1. Names, position descriptions, summaries of training, experience and location of major personnel engaged in the study must be obtained. It is also necessary to examine the workload of selected individuals to determine if they actually had time to accomplish the task specified by the protocol.

2. The QAU for the study must be identified.
3. The QAU schedule, activities, in-process inspections, including the review of the final report and retention of records must be examined. The QAU statement on the final report must be verified.
4. Significant changes in the facilities other than those currently reported must be closely examined.
5. Any equipment used in the specific study must be examined to determine if it was standardized and calibrated prior to, during and after use in connection with the study. It must be also determined, if at the time of the study, there was equipment malfunction, the impact of the malfunction on the study and the remedial action taken.
6. SOPs contemporary for the study must be evaluated.
7. The firm's records are examined to substantiate that the protocol requirements were met and, if applicable, the occurrence and types of diseases and clinical observations prior to and during the study must be examined.
8. Any significant changes in test and control article handling from those currently reported are examined.
9. A copy of the protocol is obtained by the team and checked to determine compliance with §58.120 of the regulations. It must be determined that protocol changes are properly authorized.
10. A copy of the final study report and copies of interim reports with any amendments must be obtained to determine:

   a. Whether or not the final report corresponds with the protocol and describes any subsequent changes in the protocol.
   b. Whether or not the final report accurately reflects the raw data and observations.
   c. Whether or not the final report is appropriately signed and dated and conforms to the requirements of §58.185.

SAMPLE COLLECTION

FDA investigators have the authority to collect samples as described under the Compliance Program 7348.808. Samples of a test article, the carrier, the control article or test and control article mixtures may be selected and sent to FDA laboratories to determine the identity, strength, potency, purity, composition or other characteristics which will accurately define the collected sample. In fact, even physical samples such as wet tissues, tissue blocks and slides may be collected. When the field investigator collects a sample of any chemical substance, he will also collect a copy of the methodology from the sponsor of the testing facility. The copy of the methodology will be sent to the FDA laboratory selected to perform the sample analysis.

PRESENTATION OF THE ESTABLISHMENT INSPECTION REPORT

Before concluding a GLP inspection, FDA officials meet with appropriate laboratory personnel to discuss any observed deviations from GLPs. If there are no departures from the GLP regulations, the facility representatives are so informed during the exit interview and no documentation is given to the firm. If significant deficiencies are found, the laboratory will be presented with a form FDA 483, Inspectional Observations. This form lists the deviations from the GLP regulations as observed by the FDA investigational team during the inspection. When the FDA 483 is issued during the exit interview, the representatives of the laboratory have an opportunity to discuss the statements made therein. The form may be altered or changed as a result of the exit interview discussions. The final version of the FDA 483 when issued at the end of the on-site phase of the inspection becomes immediately available under the Freedom of Information Act. As in every inspection performed under the auspices of the Act, an Establishment Inspection Report, reflecting all the findings and discussions, is prepared by the lead investigator. The report, unlike the FDA 483, is not available for release to freedom of information requests until all action on the EIR file has been completed.

PREPARATION OF THE ESTABLISHMENT INSPECTION REPORT

The lead investigator is responsible for the preparation of the EIR. However, other members of the inspection team may be called upon to participate in its preparation, particularly in supplying specialized scientific or technical information.   The field investigator and the supervisor at the district office will tentatively classify the completed EIR under one of the following three categories: NAI, VAI or OAI.

THE CLASSIFICATION PROCESS

After report preparation and establishing a proposed classification, the EIR with all its attachments and exhibits is sent to FDA headquarters. The Centers' GLP pharmacologists evaluate the EIR and make the final classification of the inspections assigned by that Center.

NAI signifies "No Action Indicated." This means that the laboratory is essentially in compliance with the GLP regulations. Ordinarily, the inspected facility receives no further correspondence from the Agency concerning the inspection and reinspection is scheduled on a routine basis.

Prior to December 1993, the classifications VAI-1, VAI-2 and VAI-3 were used to characterize the GLP compliance of an inspected facility. The term VAI is the acronym for "Voluntary Action Indicated." The numerals 1, 2 and 3 formerly indicated degrees of failure to comply with the GLP regulations. VAI-1 meant that the violations were minor and may have been corrected before the inspections was concluded. VAI-2 indicated that minor procedural deficiencies were found, which did not threaten to compromise the validity of any studies done under those circumstances. GLP inspections that were formerly classified VAI-1 and 2 are now termed VAI.

GLP violations that compromised, or potentially compromised, the scientific, and, hence, the regulatory merit of a nonclinical toxicity study were classified VAI-3. Those inspections that prior to December 1993 were VAI-3 are now classified as OAI.

The OAI, "Official Action Indicated," classification has the most serious impact. In such a case, the Center of the Agency responsible for the appraisal of the test article in question is contacted.   A recommendation is made by the GLP staff to the new drug approval pharmacologists that the study, or studies, classified OAI should not be used in support of a research or marketing permit, such as an IND or NDA.

In some circumstances, a more severe regulatory and/or administrative sanction is considered necessary by the Agency to achieve correction of the violative conditions. For instance, two or more OAI classifications, indicating that the laboratory is seriously out of compliance, could result in the disqualification of the laboratory (Title 21, Code of Federal Regulations, Part 58, Subpart K).

OAI classifications would be considered when any one or more of the following exist:

1. Quality assurance is poor or non-existent.
2. The test article and its dosage forms have not been characterized as required by Parts 58.105 and 58.113.
3. A study or studies that must comply with the GLPs have not been listed on the master schedule.
4. Numerous less serious GLP deviations that persist over two or more inspection periods. This suggests that the laboratory is out of control.

In the case of OAI classifications, a directed re-inspection will normally be assigned on a schedule determined by the Center initiating the investigations.

GLP INSPECTIONS ABROAD

It may rightfully be said that all this information is interesting in terms of laboratories in the United States, but what about GLP inspections abroad, what steps has the Agency taken in this direction? When the FDA developed the GLP regulations and its laboratory inspection program in 1976, the planners were preoccupied with domestic laboratories. A survey of investigational drug submissions completed since then found that approximately 42% of the safety studies submitted had been conducted abroad. This convinced the FDA that safety studies conducted in foreign laboratories would have to be addressed.

While the laws of this country require that the safety of foods, chemicals and drugs be demonstrated by well-controlled studies, the authority of the FDA cannot be exercised beyond the borders of the United States. The Agency, concluding that the best means of satisfying the law, would be to physically observe the operations and practices of the laboratory, took it a step further by announcing that a refusal to permit such an observation would result in the nonacceptance by the FDA of the uninspected data. This standard applies to all laboratories, both foreign and domestic, government as well as commercial, which conduct studies intended for submission to the FDA.

Since beginning inspections of foreign laboratories in 1977, the Agency has visited about half of the approximately 110 foreign laboratories that have conducted studies that have been submitted to the FDA. The FDA has inspected laboratories in most European countries as well as in Japan and Australia. Because these inspections are relatively expensive, FDA focus on international inspections is directed to laboratories that are frequent contributors of critical studies to research or marketing applications. To this point in time, the FDA's foreign GLP inspection teams have found excellent cooperation extended by the foreign laboratories. There have been no refusals to inspect and the quality of the studies audited is no better or no worse than the quality of similar studies conducted in the United States. Mainly, the GLP problems were inadequate SOPs, discrepancies between raw data and the final report, undocumented protocol changes and improper correction of recorded data.

As already mentioned, foreign GLP inspections are extremely expensive. To avoid incurring these costs, the FDA has made bilateral agreements with the drug regulatory agencies in several other countries. These bilateral agreements are developed progressively in two phases.

Phase I commits the drug regulatory agency of each country to establish a GLP program, provide for joint inspections and to share information and consultation. Once a GLP program has been established, an assessment is then made of the program to establish comparability between the inspection methods used by the foreign regulatory agency and those used by the FDA. Phase I agreements are presently active with Sweden, Japan and Canada.

The Phase II agreement, when reached by participating countries, affords reciprocal recognition of each country's program and provides for mutual acceptance of data and exchange of inspectional findings. At the present time, the U.S. Food and Drug Administration has Phase II agreements with Switzerland, Italy, France, Germany and the Netherlands.

THE RESULTS OF GLP INSPECTIONS BY THE CENTER FOR DRUG EVALUATION AND RESEARCH

Having discussed the inspection processes of FDA as far as GLP regulations are concerned, the question may then be asked: what are the recent and historic results of the FDA's toxicology laboratory monitoring program?

For Fiscal Year 1993, 47% of the 80 GLP inspections classified as involving human drugs resulted in NAI classifications.   Of the 80 inspections, 29% (23) were classified VAI-2; 9% (7) were VAI-3 (VAI-1, 2, & 3 classifications were used prior to December 3, 1993.); and 8% (6) resulted in an OAI classification. These percentages are based on the GLP inspections assigned only by the Center for Drug Evaluation and Research and do not include a small number of VAI-1 classifications where violative conditions were corrected by the laboratories prior to completion of the inspections. Since the inception of the regulations on June 20, 1979 to March 31, 1993, the Center has reviewed a total of 931 inspection reports. There were 408 inspections of sponsor labs, while the like values for contract, university, foreign and government laboratories were 435, 57, 25 and 6, respectively. The majority of the EIRs from these inspections (in execess of 80%) were classified by the Centers as either "No Action Indicated", VAI-1 or VAI-2. These results reflect favorably on the positive attitude of industry in implementing the GLPs. During the same period, 124 reports were classified VAI-3 and 47 were classified as "Official Action Indicated."

It should be noted that during this period FDA 483s were issued in 498 of the investigations, more than half of the total. It must be kept in mind, however, that the FDA 483 lists only the observed deviations from the GLP regulations; it does not prioritize the seriousness of the deviations. The significance of these observations is determined during the review and classification of the EIR at headquarters. Furthermore, the fact that an FDA 483 was not issued does not imply that the firm was in compliance. In three instances, during fiscal year 1993, when an FDA 483 was not issued, the Agency sent a letter based on the Center's evaluation of the inspection report alone. It had been concluded, based on the EIR, that although no FDA 483 was issued, the findings in the report were important enough to be communicated to the laboratory.

Of the 931 inspections classified, information letters were sent 395 times, notices of adverse findings letters were sent 106 times and letters stating study rejection were sent 49 times.

LABORATORY GLP COMPLIANCE RATINGS

Some may question how the inspected laboratories rated in terms of compliance to each of the 141 operational provisions of the GLPs. The information accumulated from the Center for Drug Evaluation and Research indicates that 66% of the inspected laboratories were cited for one or more deviations from these provisions. The most significant departures from the GLPs were: (1) final reports did not conform to the raw data; (2) improper correction of the raw data; (3) protocol revisions were implemented without amending the protocols; (4) the absence of required SOPs and the failure to amend SOPs when necessary; and, (5) the master schedule sheets and the protocols did not contain the information required by the regulations.

FDA investigators found generally acceptable performances in the archival and record retention areas as well as in the area of the physical facilities which are associated with animal care and laboratory operations. The lack of findings in these areas is encouraging, since it may be recalled that a major problem that precipitated FDA's concern for the quality and integrity of safety data was in the area of raw data retention.

GLP IMPACT AND NEW DRUG EVALUATION

With all of the foregoing information on the GLP inspection procedures and the statistical evaluation of the completed inspections, one must still ask "How have the GLP inspections impacted on new drug evaluation?"

First, the people responsible for FDA's bioresearch monitoring program are encouraged by the results of the GLP inspection seen in terms of industry's growing acceptance of the GLPs as a means of establishing a level of reliability for scientific testing.

Furthermore, we know that the deficiencies found by our inspections in the past year are not as severe as in recent years and the cooperation we are now receiving from laboratories during the investigations is at a higher level.

Lastly, and most important, the pharmacologists at the Agency, particularly those who are keenly aware of the conditions that existed before the GLP regulations came into effect, are in agreement that the GLPs have made the reviewer's tasks much easier, and they, the reviewers, feel more confident of the reliability of the information that comes across their desks.

Chapter 8

THE GLPs: THE CYBERNETIC NECESSITY

Sandy Weinberg, Ph.D.

Weinberg, Spelton & Sax, Inc., Boothwyn, Pennsylvania

What is the distinguishing characteristic that separates effective regulation from inefficient entanglement? And where do the current Good Laboratory Practices fall in that critical continuum?

The preceding essays have provided strong substantiation for the value and effectiveness of the Good Laboratory Practices. Evolving historically, having survived the trials of circumstance and time, the GLPs have proven to be regulatory guidelines of the best sort. They serve as templates for planning laboratory operations; as gauges of operational control and effectiveness; as criteria for evaluation; and as road maps for investigatory review.

Time and circumstance, however, do more than try principles. Like the winds challenging a new sapling on the hillside, the evolutionary forces of time eventually either bend or break a set of guidelines. Laboratory tests, approaches, applications, and expectations have steadily changed over the years. To effectively evaluate the continued value of the GLPs, we must examine the degree to which the Good Laboratory Practices exhibit critical cybernetic properties.

A switch that turns on your furnace will heat your home in winter. But to be maximally effective, and to avoid the discomfort and potential dangers of overheating, a cybernetic switch is required. A thermostat senses the temperature, and reduces or cuts off heat in response to the feedback it receives. This cybernetic property, the ability to respond and adjust to changing circumstances, makes the difference between a blast furnace and central heating. In regulation, the cybernetic ability to respond to and bend with changing circumstances makes the difference between effective regulation that encourages and promotes controlled and efficient operation, and either too lax or overly restrictive meddling that hampers appropriate laboratory operation.

In some regulatory circumstances cybernetic flexibility may be unnecessary, or valuable but noncritical. Contemporary laboratory operations are in a highly dynamic state, however, changing dramatically with every new test, every new approach, and every new product. The 1994 PITTCON meeting featured a laboratory equipment museum: the most impressive impact of the exhibit was the realization of the short lives of many of the now obsolete tools. The point was dramatic: laboratory evolution has proceeded at near revolutionary speeds.

This dynamism makes regulatory flexibility a cybernetic necessity. Even the most ambitious revision process of a formal regulation takes two to three years, allowing sufficient time for comment and response. Laboratory approaches, processes, and even missions change along a much more accelerated time scale. We can't wait for a new GLP addition to properly control new processes, or to prove the effectiveness of those approaches.

The EPA's Good Automated Laboratory Practices (GALPs) illustrate two aspects of the problem. Developing the GALPs was a relatively simple if time consuming process. Over a three year period, the GLPs were expanded and interpreted to more specifically define the environment of a highly automated lab. The power of the original GLPs made that process relatively easy: the Good Laboratory Practices are flexible, so that applying them to an automated environment is more a matter of interpreting than of reformulating new control concepts. The GALPs serve as evidence of the cybernetic responsiveness of the Good Laboratory Practices.

But the rejection of the Good Automated Laboratory Practices by upper level FDA managers (they have been widely adopted as an informal tool by field investigators) is even more significant. The GALPs are an EPA guideline, but the disparagement of them by FDA spokespersons has not been based on petty inter-agency rivalries. Rather, there is fundamental and appropriate realization that while EPA labs may be established for and follow relatively unchanging test methods, the labs under FDA responsibility are much more varying and dynamic. While is may be reasonable for the EPA to offer a formalized and more narrowly structured GLP interpretation for the current automated environmental lab, the power of the GLPs for FDA labs is the flexibility that comes from avoiding that kind of situational specificity. FDA field personnel may find the GALPs to be a useful tool, and FDA labs may find the GALPs to be an excellent informal guideline for eventual GLP compliance. But the variety of FDA lab applications, and the changing nature of those laboratory approaches, makes situational specific Good Laboratory Practices an inappropriate requirement.

To be useful in a dynamic environment, the GLPs must be cybernetic. There must be a rigidity of principle of control in an interpretational flexibility of application to allow an interpretive response to situational feedback. This cybernetic characteristic is present in the GLPs, providing the resiliency and guidance the preceding essays have demonstrated. In this summary of that cybernetic flexibility an examination of five changing laboratory circumstances will identify the future of the GLPs, including the need for revisions of some detail while upholding the strengths of the whole of that remarkable and effective guideline for Good Laboratory Practices.

THE RISING RISK

FDA's Center for Medical Devices is utilizing a concept known as "hazard analysis" or "risk analysis" to assess the level of supporting evidence appropriate for widely varying devices. In simplified concept, the risk analysis determines the likelihood of a critical or injurious misadventure, the seriousness of that result, and the comparative risk of other alternative treatments, strategies, or devices.

Applying the risk analysis concept to most laboratory applications would produce a very low level of real hazard. While the results of a toxicology laboratory are certainly important, the opportunities for human review and intervention are high; the threat to humans is several steps removed; and methods (both operational and statistical) are all designed to default conservatively (to err on the side of rejection rather than acceptance).

But if we hypothetically measure the relative risk in epitomizing laboratories over the past ten years, a rising risk trend is likely to emerge. More sophisticated chromatography systems decrease the human intervention and checking process. Increasingly volatile products and more efficient distribution processes decrease the insulation time provided through human reaction, while more potent biological and biologically produced products increase the degree of risk that miscalculation represents. And while defaults remain in place, automating aspects of the manufacture, distribution, labeling, and, even the research process may cause an override of the stop points that provide protection controls.

As our laboratories grow in sophistication we become increasingly aware of how important the R&D lab, the toxicology lab, the QA and stability lab, and other research and production laboratories are in the protection of ultimate human subjects. And as we turn to the treatment and cure of increasingly serious ailments and conditions, the risks represented by ineffective laboratory controls grow astronomically. The modern lab has eliminated the romance of an accidental discovery of a penicillin contamination, to avoid the dangers as mis-interpreting a potential AIDS cure, or of mislabeling the AIDS-free testing of a human blood derivative. The result is a reduction of drama, of excitement, and of the eureka of surprise breakthroughs, replacing it instead with good, steady, credible results in quality and quantity not achievable in previous environments.

Our laboratories have learned well the first rule of medicine: above all, do no harm. But in moving toward that important precept, and allowing the breakthrough and quality that a controlled conservatism presents, laboratories have relied increasingly upon the GLPs as a guideline for professionalism. And, as a good dam allows the water to rise to a more threatening height than the unlevied river, so the increased strength of the GLP laboratory now represents a greater kinetic risk than the less reliable laboratory of fifty years ago.

Areas in which regulation, causing this phenomenon of increased expectations (and hence risk), have collapsed under the weight of the new situational pressures are common and important. Arguably, the quality control problems that have faced NASA are a result of increased requirements for good QA with rigid guidelines that fail the cybernetic test: systems are increasingly dependent upon exacting standards, while standards do not rise to those new expectations.

The GLPs have passed this test, however. Increased risk has forced more documentation and supporting evidence; has forced further application of the GLP established principles; and has encouraged tighter controls on evaluation mechanisms for compliance. But the GLPs have been effective in providing guidance for the modern toxicology lab; the emergent bioagent laboratory; the nuclear medicine stability laboratory, and most everything in between.

EMPHASIS ON VALIDATION

The increased regulatory emphasis on validation probably has two direct causes. In part, it is a tribute to the effectiveness of the GLPs and the laboratory personnel who are guided by them. As regulators are forced to spend less time on the mundane they have more energy is devote to non-immediate but still critical aspects of lab operations.

In a more fundamental sense, however, the increasing concern about validation is a result of technological innovations that place increased responsibility upon laboratory systems. In one northern California lab, a Laboratory Information Management System (LIMS) is programmed to design, conduct, and interpret quality control tests. Samples are fed into a series of chromatography, reagent reaction, and spectroscopy devices. An automated reporting system evaluates the results, and determines the disposition of the sample and the batch it represents.

While all of this automation places more responsibility on the bench chemist, who must now monitor a significantly increased volume of analyses, it moves that professional an important step away from the process. That chasm is filled with a system or systems upon which lab personnel must rely. In such a circumstance it is appropriate and necessary for regulatory and the lab supervisory personnel to carefully examine the efficacy, effectiveness, reliability, and accuracy of those systems.

Have the GLPs cybernetically responded to this increase reliance upon automation, and the validation that reliance necessitates? The answer is a clear and unambiguous "yes and no." In one sense, the GLPs are woefully inadequate in this area. Validation is barely mentioned, and left undefined. Field investigators point to vague control sections of GMPs and GLPs to justify their (appropriate) interest. The GLPs offer no real guidance for compliance.

In another sense, the lack of clarification has left open the opportunity for emergent industry consensus on appropriate validation evidences, reinforcing the concept of a self regulating industry with FDA conformational checking. As validation interests and problems have evolved over the past several years, so the opportunity for corporate SOPs on validation to change and reflect those concerns has made for the necessary flexibility. And while GALPs and other non-FDA documents have provided some specification appropriate to standardized and non-evolving environments, the lack of that straitjacketing has allowed the necessary flexibility in the rapidly changing and widely varying GLP world. The very lack of detail about validation has allowed the cybernetic response that is so critical to flexibility.

As understanding about validation has spread, emphasis upon the requirements have now shifted to the manufacturers of systems as well as the laboratory users. Increasingly, laboratory equipment systems are being registered as medical devices, placing those systems under direct regulation. Laboratories in the future will be able to concentrate upon the applications and utilization of those systems, with secure knowledge that the equipment and software is itself subject to appropriate quality controls.

The GLPs will no doubt require modification to define this laboratory/ medical device division of responsibility. But to a large extent, the Good Laboratory Practices have exhibited the cybernetic flexibility to respond to the new emphasis on validation that was unanticipated during the development of the GLP guidelines. One area that will undoubtedly be crucial for change will be a clarification of the role of the notebook in the process of data development. The purpose of the notebook was to record observations and findings. It was not intuitively obvious that data could not be totally erased, but that the original value should be retained for purposes of control and audit. This same insight needs to be extended explicitly into the automated laboratory system. While early data entry utilities used in financial applications had at least a printed audit trail available, the computerized laboratory systems could forget this approach and focus solely on accuracy of input. The need to preserve the changes and updates was not always considered. This approach of correcting and updating the safeguards is one that will be continued as the GLP process continues.

The changes that might be required in the GLPs will be focused on the needs for controlling the electronic notebook, often referred to as the "electronic signature." This concept will be discussed later in the chapter. It has, however, also been subsumed under the larger category of EID (Electronic Identification). Whatever revisions are made by the regulators, the underlying assumption that records are paper based will have to be changed. This paradigm shift will require some re-evaluation of the underlying approaches as these technological changes occur. Of particular interest is the additional shift from the bench scientist with a notebook, through the scientist with terminal to the fully automated laboratory with neither scientist nor paper record. This is the area of the robotic laboratory, which we shall now address.

THE ROBOTIC LABORATORY

The previous discussion of validation suggests that automation is a trend requiring great regulatory flexibility. The robotic ultimate of that automation, though, is a rare fringe reality. Experimental robotic laboratories are not currently cost or quality effective for normal toxicology, R&D, or quality assurance environments.

The robotic devices currently in use can move samples into a testing device, generally with success in maintaining appropriate tracking on those sample sources. But except for highly specialized devices that may, for example, conduct a variety of tests on a single blood sample, full robotic automation is too risky, too expensive, and too problem prone to be a practical reality.

The exception emerges in areas in which the risk to lab personnel is too high to permit direct human interaction with samples. Some kinds of AIDS testing, nuclear research, and extreme toxicology studies argue strongly for fully automated laboratories and lab devices. As those systems become cost effective, they will improve in accuracy (with increased investment) and, for these specialized areas, in popularity.

Consider pregnancy testing which was once a laboratory procedure involving an estimated two hours of time of a laboratory professional per twenty-five samples. While lab tests are still available and often indicated (with a much higher accuracy level than thirty years ago), a home test with the same reliability of pregnancy tests dating from the introduction of GLPs is now available for under $25, requiring no professional time at all. From sample entry to reporting these simple chemical devices are, in effect, fully robotic automated laboratories.

As self-contained robotic laboratories become more common, the GLPs will be stretched to provide effective control guidance. Classifying robotic laboratories as medical devices would, of course, help: the Safe Medical Device Act will fill in the gaps left by limitations in the Good Laboratory Practices. But the cybernetic capabilities of the GLPs will be stretched too far to provide real assistance without modification or Medical Device supplementation.

ELECTRONIC SIGNATURES

Current regulations, in the GLP and elsewhere, require human signatures in a variety of places as evidence of review and approval. Advances in the security and control of documents, coupled with economic pressures to minimize paperwork, have led to pressures to accept so-called "electronic signatures" as an alternative.

In theory, an electronic signature can provide the same secure evidence as a human signature when three provisions are maintained:

First, there must be evidence that approval was given by a unique responsible person. This is generally met by restricting access to personal passwords, which can be traced to prove that a given individual appended an approval statement to specified document.

Second, there must be secure dating of the document, to assure that the approval indication is tied to a specific document version. Often this requirement is fulfilled electronically through a link to the internal system clock.

Finally, evidence that a document is unchanged after signature is required. Such evidence can be provided through a "read only" lock on the document, or through a "footprinting" process which stores an original document (or numerical summary value for the document) and compares the current version with the original through an electronic process.

If these three conditions are met, an electronic document has the same credibility as a manually signed document. A study committee of industry and regulatory sources has agreed upon procedures and standards, and current expectations call for a 1995 release of a formal announcement accepting electronic signatures.

While the GLPs can accept electronic signatures in spirit (within the three control provisions identified), the legalities of the situation require some wording revision or addendum, which is apparently being reviewed. The GLPs have the cybernetic flexibility to identify the changes needed, but will require formal adjustment to adapt to those changes.

INTENSIFICATION

Pharmaceutical laboratories have been forced in concentrate over time on smaller and smaller quantitative differences. This shift to an intensification of examination has been fueled in part by continually greater scientific sophistication; in part by a shift to the development of treatments and cures for viral rather than bacterial conditions; and, in part by increased technological improvements that allow accurate measurements in increasingly small units (e.g., the identification of smaller tumors).

As laboratories concentrate upon increasingly intensified questions, will be GLP cybernetically adapt without need for significant revision? The answer to this possibly most important question is a clear affirmative.

The GLPs are statements of principle. They are the crude first stage focusing controls on a microscope: the fine tuning is left to the internal Standard Operating Procedures of the laboratory. These SOPs have the same force of law as the Good Laboratory Practices: in effect, every lab is regulated by its own SOP-GLP set, a series of internal procedures that must be compliant with, but further expand and define, the GLPs.

As principles, the GLPs have proved to be an effective framework. As the work of regulated labs intensifies, more detailed, perhaps more rigorous SOPs will be required. But these SOPS will generally not require GLP revision. Currently, laboratories are subject to a two tier set of guidelines: the GLPs, and their own SOPs compatible with those

GLPs. As foci intensify, the same GLP principles will apply, although significant revision of the individual laboratory SOPs may be necessary.

CONCLUSION

Will the GLPs require major revision over time? Probably not. Some modifications are required to allow legal acceptance of electronic signatures; some fine tuning will be required to handle robotic environments. As high risk situations increase, further GLP principles may be added, perhaps incorporating some OSHA employee safety conditions, and expanded document and review requirements will emerge. Validation issues will continue to be clarified by joint industry and field operations, and may eventually be incorporated in the GLP standards.

Generally, however, the GLPs fare well when evaluated according to cybernetic construct. They have the flexibility to accommodate change and the clarity to direct that change. The Good Laboratory Practices have survived the test of time remarkably intact, and are likely to need little substantive revision in the future.

Both in comparison with other standards, such as the Good Clinical and Good Manufacturing Practices, and viewed independently on its own merits, the Good Laboratory Practices represent regulation at its most effective. The GLPs provide compliance guidance; focus regulatory investigation; allow situational adaptation; and provide clear principles for effective management and operations. At the same time, the GLPs incorporate cybernetic flexibility to allow increases or decreases in rigor in response to changing circumstances and requirements.

BIBLIOGRAPHY

Ackoff, Russell L., "Management Misinformation Systems." Management Science, December, 1967, pp. 147-156.

AICPA, Audit Approaches for a Computerized Inventory System, American Institute of Certified Public Accountants, New York, NY, 1980.

AICPA, Audit Considerations in Electronic Fund Transfer Systems, American Institute of Certified Public Accountants, New York, NY, 1979.

AICPA, Guidelines to Assess Computerized General Ledger and Financial Reporting Systems for Use in CPA Firms, American Institute of Certified Public Accountants, New York, NY, 1979.

AICPA, "Management, Control and Audit of Advanced EDP Systems," American Institute of Certified Public Accountants, New York, NY, 1977.

AICPA, The Auditor's Study and Evaluation of Internal Control in EDP Systems, American Institute of Certified Public Accountants, New York, NY, 1977.

AICPA, Statement on Auditing Standards No. 3, The Effects of EDP on the Auditor's Study and Evaluation of Internal Control, American Institute of Certified Public Accountants, New York, NY, 1974.

Alter, Steven, *Decision Support Systems: Current Practice and Continuing Challenges*. Addison-Wesley Publishing Company, Reading, MA, 1980.

Anon (May 1989), "Products-Information Management," *Laboratory Practice*, *38*(5):87-91.

Andersen, Arthur, "A Guide for Studying and Evaluating Internal Controls," Arthur Andersen and Co., Chicago, IL, 1978.

Anderson, C.A., "Approach to Data Processing Auditing," *The Interpreter*, Insurance Accounting and Statistical Association, Durham, NC, April 1975, pp. 23-26.

Arthus, L.J., *Measuring Programmer Productivity and Software Quality*, John Wiley & Sons, New York, NY, 1985.

Aron, Joel, The Program Development Process, Part II: The Programming Team, Addison-Wesley, Reading, MA, 1983.

Auditing Standards and Procedures. Committee on Auditing Procedure Statement No. 33. American Institute of Certified Public Accountants, New York, NY, 1963.

Auerbach Staff, "What Every Auditor Should Know About DP," Data Processing Management Service, Auerbach Publishing Co., New York, NY, Portfolio 3-09-03, 1975, p.14.

Automated Laboratory Standards: Current Automated Laboratory Data Management Practices, Final, June 1990.

Automated Laboratory Standards: Evaluation of the Standards and Procedures Used in Automated Clinical Laboratories, Draft, May 1990.

Automated Laboratory Standards : Evaluation of the Use of Automated Financial System Procedures, Final, June 1990.

Automated Laboratory Standards: Good Laboratory Practices for EPA Programs, Draft, June 1990.

Automated Laboratory Standards: Survey of Current Automated Technology, Final, June 1990.

Baird, Lindsay L., "Identifying Computer Vulnerability," *Data Management*, June 1974, pp. 14-17.

Bar-Hava, Ne., "Training EDP Auditors,"*Information and Management*, No. 4, 1981, pp. 30-42.

Barrett, M.J., "Education for Internal Auditors," *EDP Auditor*, Spring 1981, pp. 11-20.

Basden, A., and E.M. Clark (1980). Data Integrity in a General Practice Computer System (CLINICS), *International Journal of Bio-Medical Computing 11*:511-519.

Bennett, John L., ed., *Building Decision Support Systems.* Addison-Wesley Publishing Company, Reading, MA, 1983.

Berkeley, Peter E., *Computer Operations Training*, Van Nostrand Reinhold Company, New York, NY.

Bequai, August, *How to Prevent Computer Crime*, John Wiley & Sons, New York, NY.

Biggerstaff, Ted. J., and Alan J. Perlis, eds., *Software Reusability*. Vol. 2, Applications and Experience. ACM Press/Addison-Wesley, New York, NY, 1989.

Black, Henry C., *Black's Law Dictionary*, Revised Fourth Edition, West Publishing Co., St. Paul, MN, 1968.

Board of Governors of the Federal Reserve System, "Electronic Fund Transfers," *Regulation E* (12 CFR Part 205), Effective March 30, 1979 (as amended effective May 10, 1980.)

Boehm, B.W., *Software Engineering Economics*, Prentice-Hall, Englewood Cliffs, NJ, 1982.

Bonczek, Robert H., Clyde W. Holsapple, and Andrew B. Whinston, Foundations of Decision Support Systems. Academic Press, Inc., New York, NY, 1981.

Brooks, Fred, *The Mythical Man-Month*, Addison-Wesley, Reading, MA, 1975.

Bronstein, Robert J., "The Concept of a Validation Plan," *Drug Information Journal*, Vol. 20, 1986, pp. 37-42.

Brown, Elizabeth H., "Procedures and their Documentation for a LIMS in a Regulated Environment," pp. 346-358 in R.D. McDowall, ed. *Laboratory Information Management Systems*, Sigma Press, Wilmslow, U.K., 1987.

Buxton, J.M., P. Naur, and B. Randell, Software Engineering, Concepts and Techniques, Petrocelli/Charter, New York, NY, 1976.

Callahan, John J., *Needed: Professional Management in Data Processing*, Prentice-Hall, Englewood Cliffs, NJ, 1983.

Canadian Institute of Chartered Accountants Committee, "Competence and Professional Development in EDP for the CA," *CA Magazine*, Toronto, Canada, September 1974, pp. 26-70.

Canning, Richard G., "Computer Security: Backup and Recovery Methods." EDP Analyzer, January 1972.

Canning, Richard G., "That Maintenance 'Iceberg'" *EDP Analyzer*, October 1972.

Canning, Richard G., "Project Management Systems," *EDP Analyzer*, September 1976.

Caputo, C.A., "Managing the EDP Audit Function," *COM-SAC, Computer Security, Auditing and Controls*, Vol. 8, No. 1, January 1981, pp. A-1 to A-8.

Card, David N., and Robert L. Glass, *Measuring Software Design Quality*, Prentice Hall, Englewood Cliffs, NJ, 1990.

Casti, John L, Paradigms Lost. William Morrow, New York, NY, 1989.

Chapman, M., "Audit and Control in Data Base/IMS Environment," *COM-SAC, Computer Security, Auditing and Controls*, Vol. 6, No. 2, July 1979, pp. A-9 to A-14.

Charette, Robert N., *Software Engineering Risk Analysis and Management*, McGraw-Hill, New York, NY, 1989.

Cheney, P.H., "Educating the Computer Audit Specialist," *The EDP Auditor*, Fall 1980, pp. 9-15.

CICA, Computer Audit Guidelines, Canadian Institute of Chartered Accountants, Toronto, Ontario, Canada, 1975.

CICA, Computer Control Guidelines, Canadian Institute of Chartered Accountants, Toronto, Ontario, Canada, 1971.

Clary, J.B., and R.A. Sacane, "Self-testing Computers," *IEEE Computer*, October 1979, pp. 45-59.

Clinical Laboratory Improvement Act of 1967, P.L. 90-174, December 5, 1967.

Clinical Laboratory Improvement Amendments of 1988, P.L. 100-578, October 31, 1988.

College of American Pathologists, *Standards for Laboratory Accreditation*, Commission on Laboratory Accreditation, College of American Pathologists, Skokie, IL, 1988.

Connor, J.E., and B.H. De Vos, "Guide to Accounting Controls-Establishing, Evaluating and Monitoring Control Systems," Warren, Gorham & Lamont, Inc., Boston, MA, 1980.

Cooke, John E., and Donald H. Drury. Management Planning and Control of Information Systems. Society of Management Accountants of Canada, Hamilton, Ontario, April 1980.

Data Acquisition Telecommunications Local Area Networks, *1983 Data Book*. Advanced Micro Devices, Inc., Sunnyvale, CA, 1983.

Datapro Research, *Datapro Reports on Information Security*, McGraw-Hill, Inc., Delran, NJ, 1989.

Davis, G., "Ensuring On-Line System Integrity Using Parallel Simulation on a Continuing Basis," *COM-SAC, Computer Security, Auditing and Controls*, Vol.7, No. 2, July 1978, pp. A-9 to A-11.

Davis, Stanley M., *Future Perfect*, Addison-Wesley, Reading, MA, 1987.

Davis, Randal, Bruce Buchanan, and Edward Shortliffe, "Production Rules as a Representation for a Knowledge-Based Consultation Program," Artificial Intelligence, Vol. 8, 1977, pp. 15-45.

Dearden, John, "MIS is a Mirage," *Harvard Business Review*, January-February 1972, pp. 90-99.

Department of Trasportation, *Federal Register*, Procedures for Transportation Workplace Drug Testing Programs; Final Rule. Vol. 54, No. 230, December 1, 1989, 49854-49884.

Dessy, Raymond E., "The Electronic Laboratory," American Chemical Society, Washington, DC, 1985.

DeMarco, Thomas, *Controlling Software Projects*, Yourdon Press, New York, NY, 1982.

DeMarco, Thomas, *Structured Analysis and System Specification*, Yourden Press/Prentice Hall, Englewood Cliffs, NJ, 1978.

DeMarco, Thomas, and Tim Lister, *Software State-of-the-Art: Selected Papers*, Dorset House, New York, NY, 1991.

Dice, Barry, Operations Manager, Telephone Interview, Hyattsville, MD, April 25, 1990.

Dijkstra, Edsger, *A Discipline of Programming*, Prentice Hall, Englewood Cliffs, NJ, 1976.

DOD-STD-2167A. U.S. Department of Defense, Defense System Software Development, February 29, 1988.

Dorricott, K.O., "Organizing a Computer Audit Specialist," *CA Magazine*, May 1979, pp. 66-68.

Drug Information Association, "Computerized Data Systems for Nonclinical Safety Assessment: Current Concepts and Quality Assurance," Drug Information Allocation, Maple Glen, PA, 1980.

Dunn, Robert, *Software Defect Removal*, McGraw-Hill, New York, NY, 1984.

Dunn, Robert, and Richard Ullman, *Quality Assurance for Computer Software*, McGraw-Hill, New York, NY, 1982.

EDP Auditors Foundation. Control Objectives-1980, EDP Auditors Foundation, Streamwood, IL, 1980.

EIA Standard RS-422. Electrical Characteristics of Balanced Voltage Digital Interface Circuits, Electronic Industries Association, Engineering Department, Washington, DC, 1975.

EIA Standard RS-232-C. Interface Between Data Terminal Equipment and Data Communication Equipment Employing Serial Binary Data

Interchange, Electronic Industries Association, Engineering Department, Washington, DC, 1969.

Electronic Fund Transfer Act. 15 USC sec. 1693 *et. seq.*, 1979.

Emens, K.L., "A Survey of Internal EDP Audit Activity Among EDPAA Member Companies," *EDP Auditor*, Fall 1976, pp. 11-17.

Enger, Norman L., and Howerton, Paul W., *Computer Security*, AMACOM, New York, NY.

Error Detecting and Correcting Codes, Application Note AP-46, Intel Corporation, Santa Clara, CA, 1979.

Essentials of Data Communications, Tektronix Inc., Beaverton, OR, 1978.

Figarole, Paul L., "Computer Software Validation Techniques," DIA Conference on Computer Validation, January 21-23, 1985.

Fisher, Royal P., *Information Systems Security*, Prentice-Hall, Inc., Englewood Cliffs, NJ.

Forsyth, A., "An Approach to Audit an On-Line System," *COM-SAC, Computer Security, Auditing and Controls*, Vol. 7, No.1, January 1980, pp. A-1 to A-4.

Frank, E., "Integrating Reviews of EDP Systems with Regular Audit Project," *The EDP Auditor*, Summer 1974, pp. 10-11.

Freedman, Daniel, and Gerald Weinberg. *Handbook of Walkthroughs, Inspections and Technical Reviews*, Little, Brown, Boston, MA, 1982.

Gallegos, Frederick, and Doug Bieber. "What Every Auditor Should Know about Computer Information Systems," available as Accession Number 130454 from the General Accounting Office (GOA) and reprinted from p. 1-11 in EDP *Auditing*, Auerbach Publishers, Inc., New York, NY, 1986.

Gane, Chris, and Trish Sarson, Structured Systems Analysis: Tools and Techniques. Improved System Technologies, New York, NY, 1977.

Gardner, Elizabeth (1989a), Computer Dilemma: Clinical Access vs. Confidentiality, *Modern Healthcare* (November 3), p. 32-42.

Gardner, Elizabeth (1989b), Secure Passwords and Audit Trails (Sidebar), *Modern Healthcare* (November 3), p. 33.

Gardner, Elizabeth (1989c), System Assigns Passwords, Beeps at Security Breaches (Sidebar), *Modern Healthcare* (November 3), p. 34.

Gardner, Elizabeth (1989d), System Opens Access to Physicions, Restricts it to Others (Sidebar), *Modern Healthcare* (November 3), p. 38.

Gardner, Elizabeth (1989e), 'Borrowed' Passwords Borrow Trouble (Sidebar), *Modern Healthcare* (November 3), p. 42.

Gardner, Elizabeth (1989f), Recording Results of AIDS Tests can be a Balancing Act (Sidebar), *Modern Healthcare* (November 3), p. 40.

Garwood, R.M., "FDA's Viewpoint on Inspection of Computer Systems," Course notes from "Computers in Process Control: FDA Course," March 19-23, 1984.

Gild, Tom., *Principles of Software Engineering Management*, Addison-Wesley, Reading, MA, 1988.

Gilhooley, Ian A., "Improving the Relationship Between Internal and External Auditors," *Journal of Accounting and EDP*, Vol. 1, No. 1, Spring 1985, pp. 4-9.

Glover, Donald E., Rovert G. Hall, Arthur W. Coston, and Richard J. Trilling (1982), "Validation of Data Obtained During Exposure of Human Volunteers to Air Pollutants," *Computers and Biomedical Research* 15(3):240-249.

Good Automated Laboratory Practices (GALPs), December 28, 1990.

Goren, Leonard J., "Computer System Validation, Part II," *BioPharm.* February 1989, pp. 38-42.

Grimes, J. and E. A. Gentile, "Maintaining International Integrity of ON-Line Data Bases," EDPACS Newsletter, February 1977, pp. 1-14.

Guynes, Steve, "EFTS Impact on Computer Security," *Computers & Security*, Vol. 2, No. 1, 1983, pp. 73-77.

Halper, Stanley, D., Davis, Glen, C., O'Neil-Dunne, Jarlath, P., and Pfau, Pamela, R., *Handbook of EDP Auditing*, "Testing Techniques for Computer-Based Systems," 1985:28 pp. 1-26

Halstead, Maurice, *Elements of Software Science.* Elsevier, New York, NY, 1977.

Hartwig, Frederick, and Brian E. Dearing, Explanatory Data Analysis. Sage Publications, Beverly Hills, CA, 1979.

Hatley, Derek, and Imtiaz Pirbhai, *Strategies for Real-Time System Specification*, Dorset House, New York, NY, 1987.

Hayes, John R., The Complete Problem Solver, The Franklin Institute Press, Philadelphia, PA, 1981.

Hayes-Roth, Frederick, Donald A. Waterman, and Douglas B. Lenat, *Building Expert Systems.* Addison-Wesley Publishing Company, Reading, MA, 1983.

Heidorn, G.E., "Automatic Programming through Natural Language Dialogue: A Survey," IBM Journal of Research and Development, Vol. 20, No. 4, July 1976, pp. 302-313.

Hetzel, William, *The Complete Guide to Software Testing*, 2nd Ed. QED Information Sciences, Wellesley, MA, 1988.

Highland, Harold Joseph, *Protecting Your Computer System*, John Wiley & Sons, Inc., New York, NY.

Hirsch, Allen, F., *Good Laboratory Practice Regulations*, Marcel Dekker, Inc, New York, NY, 1989.

Hoaglin, David C., Frederick Mosteller, and John W. Tukey, *Understanding Robust and Exploratory Data Analysis*, John Wiley and Sons, New York, NY, 1983.

Hopper, E.L., "Staffing of EDP Auditors on the Internal Audit Staff," *The Interpreter*, August 1975, pp. 24-26.

Horwitz, G., "Needed: A Computer Audit Philosophy," *Journal of Accountancy*, April 1976, pp. 69-72.

Hubbert, J., "Data Base Concepts," *The EDP Auditor*, Spring 1980.

Hulme, K. and M.E. Aiken, "The Normative Approach to Internal Control Evaluation of On-Line/Real-Time Systems," The Chartered Accountant in Australia, July 1976, pp. 7-16.

Humphrey, Watts S., *Managing the Software Process*, Addison-Wesley, Reading, MA, 1989.

Hunter, Ronald P., *Automated Process Control Systems Concepts and Hardware*, Prentice-Hall, Englewood Cliffs, NJ, 1978.

IIA Staff, "Establishing the Internal Audit Function," Institute of Internal Auditors, Altamonte Springs, FL, 1974.

IIA Staff, "Hatching the EDP Audit Function," Institute of Internal Auditors, Altamonte Springs, FL, 1975.

IIA, "How to Acquire and Use Generalized Audit Software," The Institute of Internal Auditors, Altamonte Springs, FL, 1979.

IIA, "Systems Auditability and Control-Audit Practices," The Institute of Internal Auditors, Altamonte Springs, FL, 1977.

IIA, "Systems Auditability and Control-Control Practices," The Institute of Internal Auditors, Altamonte Springs, FL, 1977.

IIA, "Systems Auditability and Control-Executive Report," The Institute of Internal Auditors, Altamonte Springs, FL, 1977.

Jackson, Michael, *System Development*, Prentice Hall, Englewood Cliffs, NJ, 1983.

Jackson, Michael, *Principles of Program Design*, Academic Press, New York, NY, 1975.

Johnson, Curtis D., *Process Control Instrumentation Technology*, 2nd Ed., John Wiley & Sons, New York, NY, 1982.

Keen, Peter G.W., "'Interactive' Computer Systems for Managers: A Modest Proposal," *Sloan Management Review*, Fall, 1976, pp. 1-17.

Keen, Peter G.W., and Michael S. Scott Morton. *Decision Support Systems: An Organizational Perspective*, Addison-Wesley Publishing Company, Reading, MA, 1978.

Kernighan, Brian, and P.J. Plauger, *Software Tools*, Addison-Wesley, Reading, MA, 1976.

Kull, David, "Demystifying Ergonomics," *Computer Decisions*, September 1984.

Kuong, J.F., "A Framework for EDP Auditing," *COM-SAC, Computer Security Auditing and Controls*, Vol. 3, No. 2, 1976, pp. A-1 to A-8.

Kuong, J.F., "Advanced Tools and Techniques for Systems Auditing," Management Advisory Publications, 1978.

Kuong, J.F., "Approaches to Justifying EDP Controls and Auditability Provisions," *COM-SAC, Computer Security, Auditing and Controls*, Vol. 7, No. 2, July 1980, pp. A-1 to A-8.

Kuong, J.F., "Audit and Control of Advanced/On-Line Systems," (MAP-7), Management Advisory Publications, 1980.

Kuong, J.F., "Audit and Control of Computerized Systems" (MAP-6), Management Advisory Publications, 1979.

Kuong, J.F., "Auditor Involvement in System Development and the Need to Develop Effective, Efficient, Secure, Auditable and Controllable Systems," Keynote speech at the First Regional EDP Auditors Conference, Tel-Aviv, Israel, June 3, 1982.

Kuong, J.F., Checklists and Guidelines for Reviewing Computer Security and Installations (MAP-4), Management Advisory Publications, 1976.

Kuong, J.F., Computer Auditing and Security Manual-Operations and System Audits (MAP-5), Management Advisory Publications, 1976.

Kuong, J.F., "Computer Security, Auditing and Controls-Text and Readings," Management Advisory Publications, 1974.

Kuong, J.F., Controls for Advanced/On-line/Data-Base Systems Vols. 1 and 2, Management Advisory Publications, 1985.

Kuong, J.F., "Managing the EDP Audit Function," Paper presented at the First Regional EDP Auditors Conference, Tel-Aviv, Israel, May 29-June 3, 1982.

Kuong, J.F.,"Organizing, Managing and Controlling the EDP Auditing Function," Seminar Text, Management Advisory Publications, 1980.

Kuong, J.F.,"Organizing and Staffing for EDP Auditing," COM-SAC, Computer Security, Auditing and Controls, Vol. 2, No. 1, 1975.

Langmead, J.M. and R.V. Boos, "How Do You Train EDP Auditors?," *Management Focus*, September-October, 1978, pp. 6-11.

Laurel, Brenda, ed. *The Art of Human-Computer Interface Design*, Addison-Wesley, Reading, MA, 1990.

Litecki, C.R., and J.E. McEnroe, "EDP Audit Job Definitions: How Does Your Staff Compare?" *The Internal Auditor*, April 1981, pp. 57-61.

Macchiaverna, P.R., "Auditing Corporate Data Processing Activities," The Conference Board, Inc., New York, NY, 1980.

Mair, W.C., K. Davis, and D. Wood, "Computer Control and Audit," The Institute of Internal Auditors, Altamonte Springs, FL, 1976.

Marks, R.C., "Performance Appraisal of EDP Auditors," Speech given at the 11th Conference on Computer Auditing, Security and Control, ATC/IIA, New York, May 4-8, 1981.

Marks, William E., "Evaluating the Information Systems Staff," *Information Systems News*, December 24, 1984.

Martin J., *Accuracy, Security and Privacy in Computer Systems*, Prentice-Hall Inc., Englewood Cliffs, NJ, 1973.

Mason, Richard O., and E. Burton Swanson, *Measurement for Management Decision*, Addison-Wesley Publishing Company, Reading, MA, 1981.

Mattes, D.C., "LIMS and Good Laboratory Practice," pp. 332-345 in R.D. McDowall, ed., *Laboratory Information Management Systems*, Sigma Press, Wilmslow, UK, 1987.

McClure, Carma, *CASE Is Software Automation*, Prentice-Hall, Englewood Cliffs, NJ, 1989.

McClure, Carma, *Managing Software Development and Maintenance*, Van Nostrand Reinhold, New York, NY, 1981.

McDowall, R.D., ed., *Laboratory Information Management Systems*, Sigma Press, Wilmslow, UK, 1987.

McGuire, P.T., "EDP Auditoring-Why? How? What?," *The Internal Auditor*, June 1977, pp. 28-34.

McMenamin, Steve, and John Palmer. *Essential Systems Analysis*, Yourdin Press/Prentice Hall, Englewood Cliffs, NJ, 1984.

Megargle, Robert, (1989), "Laboratory Information Management Systems," *Analytical Chemistry*, *61*(9):612A-621A.

Merrer, Robert J., and Peter G. Berthrong (1989), "Academic LIMS: Concept and Practice," *American Laboratory 21*(3):36-45.

"Microelectronic," *Scientific American*, W.H, Freeman and Co., San Francisco, CA, 1977.

Miller, T.L., "EDP-A Matter of Definition," *The Internal Auditor*, July-August 1975, pp. 31-38.

Mills, Harlan, Richard Linger, and Alan Hevner. *Principles of Information Systems Analysis and Design*, Academic Press, New York, NY, 1986.

Morris, III, R., "The Internal Auditors and Data Processing," *The Internal Auditor*, August 1978.

Musa, John, Anthony Iannino, and Kazuhira Okumoto, *Software Reliability: Measurement, Prediction, Application*, McGraw-Hill, New York, NY, 1987.

Myers, Glenford, *The Art of Software Testing*, Wiley-Interscience, New York, NY, 1979.

Myers, Glenford, *Software Reliability*, John Wiley & Sons, New York, NY, 1976.

Myers, Glenford, *Reliable Software Through Composite Design*, Petrocelli/Charter, New York, NY, 1975.

Myers, Ware, "Build Defect-Free Software, Fagan Urges," *Computer IEEE*, August, 1990.

National Bureau of Standards, "Glossary for Computer Systems Security," U.S. Department of Commerce, FIPS PUB 39.

National Bureau of Standards, "Guidelines for Automatic Data Processing Risk Analysis," U.S. Department of Commerce, FIPS Publication 65, August 1979.

National Computer Security Center, Glossary of Computer Security, U.S. Department of Defense, NCSC-TG-004-88, Version 1, 1988.

Norris, P.N., "EDP Audit and Control - A Practioner's Viewpoint," *The EDP Auditor*, Winter 1976, pp. 8-14.

Office of Information Resources Management, "EPA LIMS Functional Specifications," Environmental Protection Agency, Washington, DC, March 1988.

Office of Information Resources Management, *EPA System Design and Development Guidance*, Vols. A, B, and C, U.S. Environmental Protection Agency, Washington, DC, 1989.

Office of Information Resources Management (1989a), *Survey of Laboratory Automated Data Management Practices*, U.S. Environmental Protection Agency, Research Triangle Park, NC, 1989.

Office of Management and Budget, *Guidance for Preparation and Submission of Security Plans for Federal Computer Systems Containing*

Sensitive Information, OMB Bulletin No. 88-16, Office of Management and Budget, Washington, DC, July 6, 1988.

Parikh, Girish, "The Politics of Software Maintenance," *Infosystems*, August 1984.

Patrick R.L., "Performance Assurance and Data Integrity Practices," National Bureau of Standards, Washington, DC, 1978.

Perry, W.E., "Auditing Computer Systems," FAIM Technical Products, Inc., Melville, NY, 1977.

Perry, W.E., "Internal Control," FAIM Technical Products, Inc., Melville, NY, 1980.

Perry, W.E., "The Making of Computer Auditor," *The Internal Auditor*, November-December 1974.

Perry, W.E., "Adding a Computer Programmer to the Audit Staff," *The Internal Auditor*, July 1974, pp. 1-7.

Perry, W.E., "Career Advancement for the EDP Auditor," *EDPACS Newsletter*, February 1975, pp. 1-6.

Perry, W.E., "Snapshot -- A Technique for Tagging and Tracing Transactions," *EDPACS Newsletter*, March 1974, pp. 1-7.

Perry, W.E., "Trends in EDP Auditing," *EDPACS Newsletter*, December 1976, pp. 1-6.

Perry, William E., *Ensuring Data Base Integrity*, John Wiley and Sons, New York, NY, 1980.

Perry, William E., "Using SMF as an Audit Tool-Accounting Information," *EDPACS Newsletter*, February 1975.

Perry, W.E. and J.F. Kuong, "Developing an Integrated Test Facility for Testing Computerized Systems" (MAP-12), Management Advisory Publications, 1979.

Perry, William E., and Donald L. Adams, "SMF -- An Untapped Audit Resource," *EDPACS Newsletter*, September 1974.

Perry, W.E., and J.F. Kuong, "Effective Computer Audit Practices Manual (ECAP)," Management Advisory Publications, Wellesley Hills, MA.

Perry, W.E. and J.F. Kuong, EDP Risk Analysis and Controls Justification, Management Advisory Publications, 1981.

Perry, W.E. and J.F. Kuong, "Generalized Computer Audit Software-Selection and Application" (MAP-14), Management Advisory Publications, 1980.

Phipps, Gail, "Practical Application of Software Testing," Computer Sciences Corporation, presented October 8, 1986.

Pinkus, Karen V. (1989), Financial Auditing and Fraud Detection: Implications for Scientific Data Audit. *Accountability in Research 1*:53-70.

Polanis, M.F., "Choosing an EDP Auditor," *Bank Administration*, January 1973, pp. 52-53.

Pressman, Roger S., *Making Software Engineering Happen*, Prentice Hall, Englewood Cliffs, NJ, 1988.

Ravden, Susannah, and Graham Johnson, *Evaluating Usability of Human-Computer Interfaces*, John Wiley & Sons, New York, NY, 1989.

Reilly, R.F. and J.A. Lee, "Developing an EDP Audit Function: A Case Study," *EDPACS Newsletter*, May 1981, pp. 1-10.

Reimann, Bernard C., and Allen D. Warren, "User-Oriented Criteria for the Selection of DDS Software," Communications of the ACM, Vol. 28, No. 2, February, 1985, pp. 166-179.

Romano, Carol A. (1987), Privacy, Confidentiality, and Security of Computerized Systems: The Nursing Responsibility, *Computers in Nursing* (May/June), pp. 99-104.

Rugg, Tom, *LANtastic*, Osborne McGraw-Hill, Berkeley, CA.

Sandowdki, C., and G. Lawler (March 1989), "A Relational Data Base Management System for LIMS," *American Laboratory 21*(3):70-79.

Savich, R.S., "The Care and Feeding of an EDP Auditor," *EDP Auditor*, Summer 1974, pp. 12-13.

Schatt, Stan, *Understanding Local Area Networks*, Third Edition, SAMS, Carmel, IN.

Schindler, Max, *Computer-Aided Software Design*, John Wiley & Sons, New York, NY, 1990.

Schneidman, A., "A Need for Auditors' Computer Education," *The CPA Journal*, New York, NY, June 1979, pp. 29-35.

Schroeder, Fredrick J. (1983), "Developments in Consumer Electronic Fund Transfers," *Federal Reserve Bulletin 69*(6):395-403.

Schulmeyer, G. Gordon, *Zero Defect Software*, McGraw-Hill, Inc., New York, NY, 1990.

Schuyler, Michael, *Now What?*, Neal-Schuman Publishers, New York, NY.

Scoma Jr., I., "Data Management," *The EDP Auditor*, May 1977, pp. 14-17.

Smith, Martin R., *Commonsense Computer Security*, McGraw-Hill Book Company, London, England, 1989.

Software Quality and Reliability, Edited by Darrel Ince, Chapman and Hall, London, England, 1991.

Sulcas, P., "Planning Timing of Computer Auditing," *The South African Chartered Accountant*, July 1975, pp. 232.

Tussing, R.T. and G.L. Hellms, "Training Computer Audit Specialists," *Journal of Accountancy*, July 1980, pp. 71-74.

United States Department of Defense, Defense System Software Development Data Item Description for the Software Design Document, DI-MCCR-80012A, GPO, Washington, DC, February 29, 1988.

U.S. Department of Health and Human Services, *Federal Register*, Mandatory Guidelines for Federal Workplace Drug Testing Programs; Final Guidelines, Vol. 53, No. 69, April 11, 1988, 11969-11989.

U.S. Department of Health and Human Services, Federal Register, Medicare, Medicaid and CLIA Programs; Final Rule with Comment Period, Vol. 55, No. 50, March 14, 1990, 9537-9610.

U.S. General Accounting Office, "Evaluating the Acquisition and Operation of Information Systems" General Accounting Office, Washington, DC, 1986.

U.S. General Accounting Office, Bibliography of GAO Documents, *ADP, IRM, & Telecommunications 1986*, General Accounting Office, Washington, DC, 1987.

Vasarhelyi, M.A., C.A. Pabst, and I. Daley, "Organizational and Career Aspects of the EDP Audit Function," *EDP Auditor*, 1980, pp. 35-43.

Weinberg, Sanford B., *System Validation Checklist*, copyrighted monograph, 1988.

Weinberg, Sanford B., *Testing Protocols for the Blood Processing Industry*, copyrighted monograph, 1989.

Weinberg, Sanford B., *System Validation Standards*, Kendall/Hunt Publishing Company, Dubuque, IA, 1990.

Wesley, Roy L., and Wanat, John A., *A Guide to Internal Loss Prevention*, Butterworth Publishers, Stoneham, MA.

Wilkins, B.J., *The Internal Auditor's Information Security Handbook*, Institute of Internal Auditors, Altamonte Springs, FL, 1979.

Willingham, J., and D.R. Carmichael, *Auditing Concepts and Methods*, McGraw-Hill, New York, NY, 1975, pp. 271-273.

Yourdon, E., *Structured Walkthroughs*, Yourdon Press, New York, NY, 1982.

INDEX